Communities, Biodiversity and Conservation of Similipal

The Editors

Prof. Hrudayanath Thatoi, M.Sc., M. Phil., Ph.D. is currently serving as a Director of the Industrial Biotechnology Research (CIBR), Siksha "O" Anusandhan, Deemed to be University, Bhubaneswar. Previously, Dr. Thatoi served as a Professor and Head of the Department of Biotechnology, at MSCB University. Dr. Thatoi has implemented several research projects funded by UGC-DAE, DST and Forest Department, Government of Odisha. Dr. Thatoi has published 25 books including two textbooks and more than 326 research papers in national and international journals and awarded two patents. He is a recipient of the Samanta Chandra Sekhar Award by DST, Government of Odisha, Prof. Harihar Pattanaik Award for environment science by Orissa Botanical Society and a Fellow of the Society of Applied Biotechnology. He has guided 20 PhD scholars. He has visited Bangkok, Nepal, and Malaysia to attend international conferences on biodiversity and biotechnology.

Prof. Hemanta Kumar Sahu, M.Sc., M. Phil., Ph.D. joined as a Lecturer in Zoology at erstwhile North Orissa University, now, Maharaja Sriram Chandra Bhanja Deo University, Baripada, Odisha, India in 2004 and has been a Professor since 2015. He is a Member of the Society of Conservation Biology, Asia Chapter and IUCN World Commission on Protected Areas (2021-25). He has over 100 publications in peer-reviewed national and international journals with h-index16 and i10 index 29. He has also authored 15 books. He has visited several countries such as the USA, Malaysia, Indonesia, Thailand and UAE to participate in training programs and conferences on biodiversity conservation. He has awarded 15 Ph. D.s and currently mentoring 4 scholars.

Dr. Cuckoo Mahapatra, M.Sc., M. Phil., Ph.D. is working as Assistant Professor of Zoology, Maharaja Sriram Chandra Bhanja Deo University (Erstwhile North Orissa University), Baripada, Odisha, India since 2016. She received the Fulbright-Nehru Pre-Doctoral and Professional Fellowship in 2010 to pursue research on salamander regeneration at the University of California Irvine, Irvine, CA, USA. She has received grants from DST-SERB, UGC, BRNS, and CSIR for working on regeneration biology, the impact of high natural background radiations and environmental contaminants on amphibians. She has around 30 publications in indexed and peer-reviewed journals including 4 books.

Dr. Sudam Charan Sahu, M.Sc., Ph.D. FIAT is working as an Assistant Professor in the Department of Botany, Maharaja Sriram Chandra Bhanja Deo University (Erstwhile North Orissa University), Baripada (Odisha), India. He has published more than 70 papers in various national and international journals with h-index 16 and i10-index 20. He has edited two books, authored three books and 16 book chapters. He is recognized as Fellow of Indian Association for Angiosperm Taxonomy. He was also awarded with DST-Young Scientist from Science & Engineering Research Board (SERB), DST, Government of India.

Communities, Biodiversity and Conservation of Similipal

– Editors –

Hrudayanath Thatoi

Hemanta Kumar Sahu

Cuckoo Mahapatra

Sudam Charan Sahu

2025

Daya Publishing House®

A Division of

Astral International Pvt. Ltd.

New Delhi – 110 002

ISBN: 978-93-5919-453-0

Published by : **Daya Publishing House®**
A Division of
Astral International Pvt. Ltd.
– ISO 9001:2015 Certified Company –
4736/23, Grond Floor, Ansari Road,
Darya Ganj, New Delhi-110 002
Ph. 011-43549197, 8130496929
e-mail: info@astralint.com
Website: www.astralint.com

Printed at : **Replika Press Pvt Ltd**

MAHARAJA SRIRAM CHANDRA BHANJA DEO UNIVERSITY
SRIRAMCHANDRA VIHAR, TAKATPUR, BARIPADA-757003, ODISHA
Phone: 06792-255127(O), 278275® e-mail ID- vconou@rediffmail.com

Foreword

It is a great pleasure to write the foreword for this remarkable book, "*Communities, Biodiversity and Conservation of Similipal*", an outcome of the IDP OHEPEE supported National Seminar on "Similipal: Communities, Biodiversity and Conservation" organized by the Centre of Similipal Studies, Maharaja Sriram Chandra Bhanja Deo University. As the Vice-Chancellor of Maharaja Sriram Chandra Bhanja Deo University, I am delighted to see our university's faculty and researchers contributing to such a pivotal publication.

The Similipal landscape, a UNESCO Biosphere Reserve, and its adjoining areas are a treasure trove of biodiversity and home to numerous indigenous communities. This book is a testament to the intricate relationships between these communities, the biodiversity they inhabit, and the conservation efforts that are essential to preserving this delicate balance.

Through rigorous research and nuanced analysis, the authors shed light on the complexities of community-led conservation, the impact of human activities on biodiversity, and the role of policy and governance in shaping the future of this precious ecosystem.

This book is a significant contribution to the fields of conservation biology, environmental studies, and community development. It will serve as a valuable resource for scholars, policymakers, and practitioners working towards the sustainable management of natural resources and the well-being of local communities.

I commend the editors and authors for their tireless efforts in bringing this project to fruition. I am confident that this book will inspire meaningful dialogue, inform evidence-based decision-making, and ultimately contribute to the conservation of the Similipal landscape for generations to come.

Once again, I congratulate the team on this outstanding publication.

Santosh Kumar Tripathy

Preface

The Similipal Biosphere Reserve is a gem in the crown of Mayurbhanj, known worldwide for its unique ecological significance and designated as a UNESCO Heritage Site. It is one of the foremost protected areas in India and holds a special place in the hearts of the people of Mayurbhanj. Similipal serves multiple roles: it is a wildlife sanctuary, a biosphere reserve, a tiger reserve, an elephant reserve, and an Important Bird Area (IBA) and a National Park. Located in the northern part of Odisha, within the Mayurbhanj district, Similipal features vast landscapes dominated by sal forests, which connect with the protected areas of Hadgarh and Kuldiha to the south, as well as the forests of Keonjhar to the west. It is the largest of the 19 protected areas in Odisha and one of the first nine tiger reserves established in 1973.

Biogeographically, Similipal is part of the Chhotanagpur Plateau, bordered by the districts of Balasore to the east and Keonjhar to the southwest. The total area of the Biosphere Reserve spans 5,569 square kilometers, with Similipal Sanctuary at its core. The region's hills rise sharply from the plains of Udala in the south and Baripada in the southeast, extending to Jashipur in the northwest, Bisoi in the north, and Thakurmunda to the west.

Similipal boasts exceptional faunal diversity compared to other protected areas in the Deccan peninsula. Recorded fauna includes 61 species of mammals, 364 species of birds, 62 species of reptiles, 20 species of amphibians, 66 species of fish, 188 species of butterflies, and numerous other invertebrates. It hosts the highest tiger population in Odisha, along with other important and conservation-dependent species such as elephants, gaurs, leopards, leopard cats, Asian small-clawed otters, stripe-necked mongooses, chousinghas, and giant squirrels.

In terms of flora, the reserve contains 1,107 indigenous species of vascular plants, including 713 species of dicotyledons from 123 families, 328 species of monocotyledons from 24 families, three species of gymnosperms, and 61 species of pteridophytes from 27 families. Furthermore, at least 274 algal taxa, 43 species of bryophytes, and 128 species of aquatic angiosperms have been reported in Similipal.

Mayurbhanj is home to several tribal communities, including the Ho, Kolha, Santhal, Bathudi, Bhumija, Mahali, Saunti, Munda, Gonda, and Pauri Bhuiyan. It also contains two primitive groups: the Erenga Khadias and Mankirdias, who inhabit Similipal and its surrounding areas. The Santhal are the largest tribal group, while the Mankirdias are the smallest, with only 103 households remaining. These tribes rely heavily on forest resources for their livelihoods.

Recently, Similipal has faced challenges from forest fires, poaching, exploitation, and deforestation. Additionally, the increasing number of villages within and around the protected area has added pressure on its resources. Government agencies and NGOs are working diligently to conserve its rich biodiversity while providing livelihood support to those dependent on Similipal. The relationship between biodiversity and tribal communities has been vital for sustenance and survival throughout history. The rivers, hills, wildlife, and forests of Similipal are deeply embedded in the local culture, and these communities depend on forest products for their livelihood. Their lifestyle fosters harmony with nature and natural resources. Thus, it is important to highlight the biodiversity and communities of Similipal to understand the traditional conservation mechanisms adopted by indigenous tribal people. This book represents the first serious attempt to explore these themes. It contains comprehensive information about Similipal, including its flora, fauna, and sustainable management practices, contributed by prominent researchers from across the country. Taxonomists, researchers, and scientists in biological fields, as well as teachers, students of biology and pharmacy, policymakers, foresters, and others engaged in the conservation, utilization, and sustainable management of the rich biodiversity in this region will greatly benefit from this book.

Hrudayanath Thatoi

Hemanta Kumar Sahu

Cuckoo Mahapatra

Sudam Charan Sahu

Contents

Chapter 1

Lessons while Shaping the Research Pathways (1979-2003) for Biodiversity Conservation and Management in Similipal

L.A.K. Singh

Former Senior Research Officer (Wildlife), Government of Odisha
e-mail: laksinghindia@gmail.com

Introduction

For success in conservation, 'research' is an integral part. For me the motivation for immediate subjects of research came from some kind of field necessity, or while pursuing an accidental observation from natural history, or a challenging task given to me for solution. Therefore, in the context of Similipal Biosphere Reserve, the research pathways that got created, the results which emerged, and the lessons gradually learnt were out of first-hand experience, unique and of historical significance.

In December 1979, I had an invited opportunity for surveying the rivers and streams in Similipal (Singh 1979) to ascertain the presence of the crocodile and suggest locations for their reintroduction. From November 1987, I got a posting in Similipal Tiger Reserve (STR), and started staying in the campus of the Mugger Project at Ramatirtha. This time it was a longer opportunity to work with *Crocodylus palustris* in the context of Similipal, and have better understanding of animals in water, on land and the upper canopy of the forest.

In order to have an insight to the animals normally visible in Similipal, I picked up the animal sighting registers maintained at four saltlicks, namely Brundaban, Chahala, Devasthali and Upper Barha Kamuda. It enabled to understand the visit patterns of elephants, with biological and management ramifications. With simple ancillary information on the movement of elephants, their corridors linking Similipal, and thus, it was possible to develop the pathway for writing the first Project Elephant document for Orissa state, that was discussed at the state level on 10 August 1989. It included the creation of elephant reserves, identification of elephant movement corridors and various mitigation and management of issues arising out of man-elephant interfaces.

Methods for Studies on Tiger Population

Began with Compliance for a Two-Page Tracking-Note

The focussed pathway for tiger (*Panthera tigris*) research began for me in 1988. Mr Mahesh Chandra Das, the Chief Conservator of Odisha (Orissa) said over tea at Chahala FRH that 'in spite of being the head of the Forest Department he was not very clear and sure about the method adopted for census of tiger'. Then he asked me to prepare a two-page note in Odia language for reference by all staff and officers, and ordered for organising the all-Orissa tiger census in 1989. That was the impetus and could gather full cooperation and collaborations, for first of all, thoroughly understanding and improving the method of pugmark tracking.

Refinements in Pugmark Tracking

My knowledge about pugmark tracking was miniscule from 1979 at Satkoshia Gorge Sanctuary, when Mr Balabhadra Prasad Das was the Wildlife Warden. Data collected from the field were sent to Mr S. R. Choudhury, the founder Field Director of STR from 1973. Mr Choudhury was also the architect of the national level first All India Tiger Census in 1972. My subsequent assignments and associations at CCBMTI, the Central Crocodile Breeding and Management Training Institute, Hyderabad, had equipped me with some secondary familiarity with pugmark tracking, aptly called 'Cooperation Census'.

Over extensive field trekking and fireside sit-outs, I had very intimate discussions with staff of STR at all levels. Most of them like Tankadhar Mohanta, Harmohan Barik, Ramesh Chandra Mohanta and Kulamani Mohanta, were founder staff of the Tiger Reserve from 1974. The discussions helped and guided me well. Immediate improvements I wanted and implemented in the existing technique were, (1) changing the season of census from summer to winter; (2) reconnaissance survey of numerous animal tracks, forest roads and jeepable roads; (3) developing and putting to use of a network of census routes initially identified to run through about 45 census beats, that got increased to 75 units by the year 2003; (4) start understanding the differences between (a) impressions of front pug abundant near water bodies from those of the hind pugmarks on various types of roads, (b) left pugmark from right pugmark as an additional parameter for individual identification, (c) identification of the sex of tiger from pugmarks, (d) distinction between the pugmark of tiger cubs and adult leopards, (e) determination of movement areas and interlinking of closeness between sets of male-female-cubs of tiger and leopard through linear profiling of pugmark parameters, (5) before census deciding about the placement and number of Pug Impression Pads (PIP) along various census routes, and (6) developing with the Working Plan Organisation, the map showing the locations of forest compartments or census units for ultimate mapping of average location of movement area of each tiger.

At the instance of project tiger directorate, non-officials were welcomed as participants during a few years from 1990s. A few participants came from outside the state and outside the country. The participants were given training about field data collection. As per their availability, they participated in the concluding discussions, experience sharing and data analysis. The 'two-page note' prepared for 1989 census got improved in many subsequent texts, explanations and illustrations. An hour-long training video was developed with staff and technical support from video technician at Baripada. A guideline was prepared at the suggestion of WWF-India, which they published and circulated in 1999, with vetting by neighbouring countries. It was revised in the year 2000 and was put in use in all neighbouring tiger-range countries. Because of simple language in English and Hindi, all staff could understand the technique, use it in the field and could make conclusions while gaining fondly ownership about tigers in their area.

Therefore, the large cat census in Similipal Tiger Reserve were carried out according to Pug-mark tracking (Choudhury 1970, 1979; Panwar 1979ab, Sawarkar 1987) with improvements and refinements made since 1989 (Sagar and Singh 1990, 1991, 1993; Singh 1995, Singh 2000a, 2000b). The results were manyfold, including the Chaturvedi Prize from Indian Forester, Forest Research Institute- Dehradun. The World Wildlife Fund - India Tiger Conservation Programme presented a medal, certificate and cash prize because of innovative and significant contributions.

Tiger Carrying Capacity predicted in 1995 from Pug-tracking

Data collected on movement areas resulted in location maps. Over the years from 1989 to 2004 the maps showed that the tigers were vacating or moving away from some areas and shifting towards the southern part of the core area (Singh 2021). Such congregations in southern Similipal were regarded as the intermediary stages through which we got indications of having reached the tiger carrying capacity as reported in Prusty and Singh (1996a). Thereafter, we predicted about the imminent trend of population decline (Srivastava and Singh 1997).

Thus, based from pugmark tracking knowledge could be gathered about the people-linked trouble areas for tiger, the tiger-linked congregations, and the warnings about approach to carrying capacity with changing characteristics of large carnivore composition in Similipal (Singh and Swain 2005). However, the warnings and predictions didn't create expected impact until camera trap results on population from 2006 onwards (Jhala *et al.,* 2018) were available through National Tiger Conservation Authority (NTCA) and Wildlife Institute of India (WII).

Table 1.1: Trend of Population of Tiger in Similipal Tiger Reserve

Year	*Male*	*Female*	*Cub*	*Total*
1972	Incomplete assessment			17
1975				30
1976	Improving coverage of survey area			46
1977	23	29	4	56
1979	22	39	4	65
1984	31	43	7	81
1986	32	51	6	89
Improvement and innovations in Pugmark Tracking aimed at minimum numbers, composition, movement area.				
1989	21	51	21	93
1990	22	52	20	94
1991	24	50	22	96
1992	24	49	22	95
1993	24	49	22	95
1995	25	47	25	97
Prusty and Singh (1996): Informed about stagnation of tiger population and approach of carrying capacity in Similipal. Srivastava and Singh (1997): Predicted year 2008 for start of declining trend in tiger population. The trend of tiger popularion was getting flattened. Human population inside Tiger Reserve was increasing.				
1997	26	48	24	98
1999	28	44	26	98
2000	29	39	29	99 (incl. 2 unknown sex)
2002	29	43	27	99
2004	28	42	31	101
2006, 2010, 2014, 2018	New 'sample census' method using camera traps and computer programmes			

Tiger Breeding Depression Predicted from 1993

The Clue from Melanistic Tiger of Podagad

Podagad is a village on the western outskirt of STR in Bhandan River valley, approachable from Jashipur. On 21 July, 1993, a young melanistic tigress got killed by Salku, a tribal boy, while giving protection to his father Surai Besra. The incident occurred close to the sugarcane field near their home. The major peculiarity in the body colour of the tiger was that the coat had a black background with tawny or white stripes on the back and underside. The colour was not normal and not black. It was melanistic. The size and behaviour of the animal, and the location of the incident in the periphery of Similipal, confirm that the young tigress was in her transient phase and inexperienced (Prusty and Singh 1996 b, c; Singh 1997, Srivastava and Singh 1998).

Figure 1.1: The First Melanistic Evidence with Salku (4th from left) of Podagad and Staff of Similipal, 21 July 1993.

Earlier to this case, on 27 July 1988 a stripe-less white tiger was sighted near Brundaban saltlick in north west Similipal. This account given in Sagar and Singh (1989) also shares three other sightings. After five years, on 21 June 1993, the body of the freshly killed melanistic tiger from Podagad was obtained. This led to further studies with assistance from WWF-India. Singh (1998, 1999) gave mapped locations for eleven sightings of melanistic tigers in STR during the period 1993-1998, and also mentioned the sighting of a melanistic tiger in Satkoshia Gorge Sanctuary in March 1997. From photographs shared by Dr J. Marcan (Singh 1998, 1999), the variations among white tigers, with dull white body and extensive dark stripes, *etc.*. appear within the white possibilities. In 1980, white tiger cubs were born to normal parents at Nandankanan Biological Park, Odisha (Roychoudhury and Acharjyo 1983). Desai and Malhotra (1992) presented a review of incidents relating to wild-sighting and captive-birth of white tigers. In 2010-2011, melanistic cubs were born in Nandankanan Biological Park, and white and melanistic cubs were born in Arignar Anna Zoological Park, Vandalur, Chennai. Within the scope through epistasis of agouti and tabby genes, a simple graphical population model got shape over the years to suggest the range of possible colour variations. Knowledge from natural history, ancient paintings (Divyabhanusinh 1986), nature-tourists, amateur photographers and simple naked-eye observers are being duly considered as contributions to natural history data (Singh 2015).

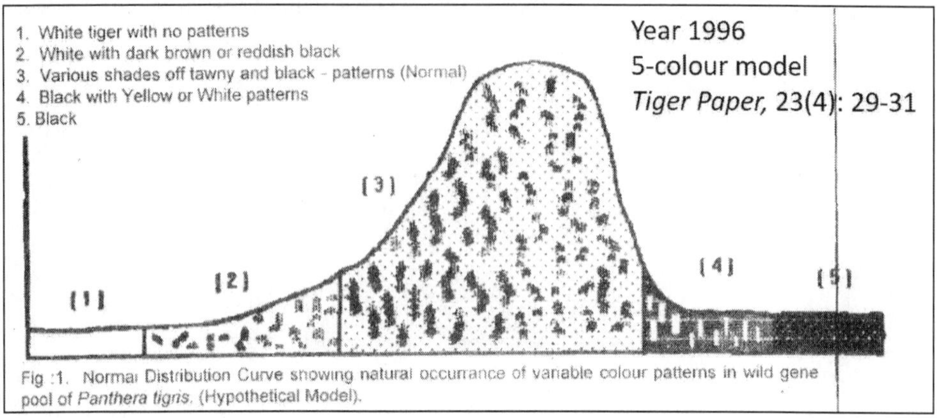

Figure 1.2: The 5-Colour Model for the Range of Colour in Tiger, *Panthera tigris*, 1996. This prototype model shows a normal distribution curve showing natural occurrence of variable colour patterns in wild gene pool of tiger, but suggesting definite occurrence of patterns in between the stages identified in Prusty and Singh (1996).

In the year 2020, photographs of a wild golden tiger from Kaziranga National Park, north-east India (Narayanan 2020) got added to the existing data. It confirmed the prediction (Singh 1997, 1999) that the variations in body colour of tiger appears to be a natural phenomenon with all natural populations of tiger. Because of the natural order of tiger biology, we fail to notice the rare and unwanted individuals in the population.

14-Colour Model Depicting Variations for *Panthera tigris* (1993-2000)

Subsequent to the case study of melanistic tiger melanistic tiger and the known instances of white tigers and the black tiger, models of colour distribution curve were developed that depicted stages of 5-colour variations in tiger. Later, developments were made of 9-colour model, 12-colour model and the range of 14 colour variations. The 14 colours that could be identified in the normal distribution curve are the (1) Snow-white or Stripe-less white, (2) reduced stripes on white background, (3) lighter or faded stripes on white tigers, (4) darker stripes on white tigers, (4) various shades of white, (5) golden or pallid, (6) normal tiger with light-yellow body, (7) normal tiger, (8) normal tiger with deep yellow body, (9) rufous, (10) brown with dark stripes, (11) brown without stripe, (12) melanistic, (13) blue, and (14) black tigers.

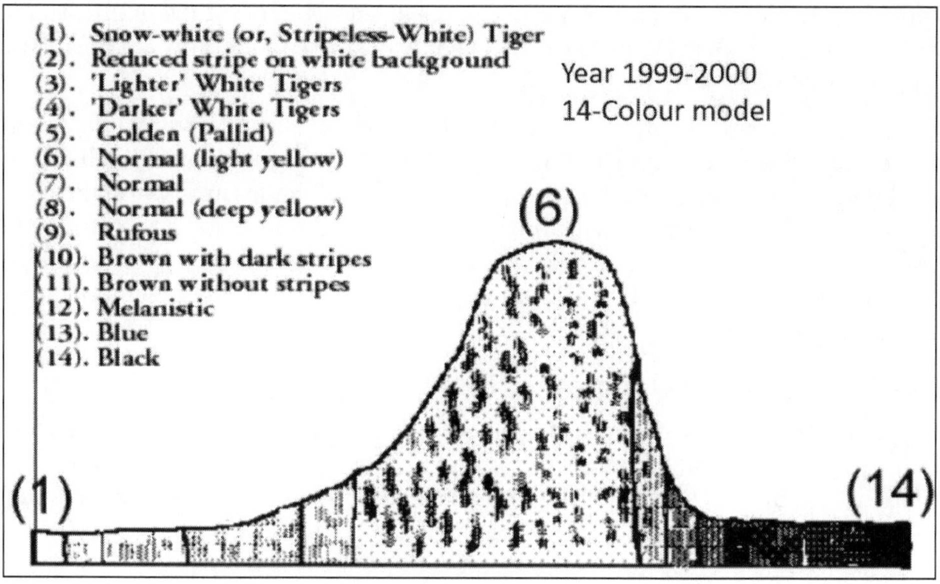

Figure 1.3: The 14-Colour Model for the Range of Colour in Tiger, *Panthera tigris,* as Described in Singh (1999). 1. Stripeless white, 2. Reduced stripe on white background, 3. White, 4. 'Darker' whites, 5. Golden (Pallid), 6. Normal (light) yellow, 7. Normal, 8. Normal (deep) yellow, 9. Rufous, 10. Brown with dark stripes, 11. Brown without stripes, 12. Melanistic. 13. Blue, 14. Black.

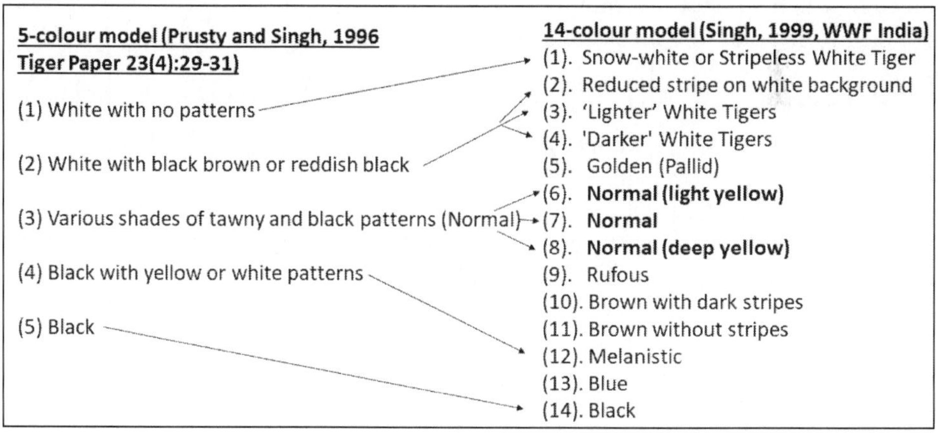

Figure 1.4: Growth of the Concept about Range of Colour Variation in *Panthera tigris.*

The 14-colour model accommodates and explains the known types of global colour variation in tiger. It also highlights the biological and conservation implications of inbreeding depression threatening the genetic integrity of tigers (Singh 1999, 2000c). After 28 years of detection of the Podagad melanistic tiger and the statements of concern for biological and conservation implications, molecular studies by Sagar *et al.* (2021) confirmed preponderance of melanistic tiger in Similipal.

In this context an overview of past researches was discussed along with the current trends of wildlife research, and the lessons for future (Singh 2014) at the erstwhile North Orissa University, Baripada. The lessons were ahead of modern times, and relate to human-induced habitat degradation and population fragmentations. The lesson is also that let's own and quickly act upon warnings from indigenous and traditional knowledge, when sophisticated equipment or technology were not available

The 14-Colour Model Applies to Leopards

Black leopard (*Panthera pardus*) is known since long (Ali 1927, van Ingen and van Ingen 1941; Divyabhanusinh 1993; Sinha 1996; Sayyed *et al.,* 2013; Sayyed and Mahabal 2015; Hedges *et al.,* 2015; Mahabal *et al.,* 2019; Wikipedia 2020; Parihar 2021; YouTube 2010, 2018a, b). Of late, melanistic and black leopard Images are seen from camera-trapping and normal photographs (Palei *et al.,* 2018, Hambali *et al.,* 2021) (Image-1). The images released by Parihar (2021) shows the photo of a melanistic leopard and one having different shades of normal body colour. The photographs of leopard with pink or golden colour have also been released in press (Dsouza 2021, TNN 2021). These leopard colours get accommodated within the 14-colour model drawn for tiger. It is expected that information on leopards of body colour other than the normal will be added in future in the database. At this time, more sightings or *ex-situ* sampling will help us to know the segregation frequency of non-normal colour expressions without inbreeding depression.

Conclusions

Using simple observations and indigenous skill, it was possible to infer about tiger carrying capacity, signs of inbreeding depression, and about the imminent declining trend of tiger population. The conclusions and predictions made for tiger population in Similipal from 1987 onwards were later proved with modern equipment like camera trap and sophisticated computer applications from 2006 onwards and molecular studies for DNA profiling in 2021. Without sophisticated equipment and laboratory techniques observations, conclusions and serious warnings on tiger population of Similipal were possible. Let's not ignore such conclusions or warnings about wildlife management.

REFERENCES

Ali S (1927) Black Leopards. Journal of Bombay Natural History Society, 31(4): 1027

Choudhury SR (1970) Let us count our tigers. Cheetal 14(2): 41-51.

Choudhury SR (1979) The Second All India Tiger Census Due in Summer 1979. Pages 300-304. In: International Symposium on tiger, India, February 22-24,

1979. Papers, proceedings and resolutions. Project Tiger, Government of India, New Delhi, pp 417.

Desai JH and Malhotra AK (1992) Publications Division, Government of India, Delhi.

Chavda D (1986) The Earliest Record of a White Tiger (*Panthera tigris*). Vol.83 (Supplement): 163-5.

Chavda D (1993) Note on Mutant leopards (*Panthera pardus*) from India. JBNHS 90: pp.88-9.

Dsouza C (2021) In a first, rare pink leopard sighted in Rajasthan (https: // timesofindia.indiatimes.com/toireporter/author-Claude-Dsouza-479256203.cms)/TNN/Updated: Nov 10, 2021, 13: 23 IST.

Hambali K, Fazli NFM, Amir A, Fauzi N, Hassin NH, Abas MA, Karim MFA and Sow AY (2021) The discovery of a melanistc Leopard *Panthera pardus delacouri* (Linnaeus, 1758) (Mammalia: Carnivora: Felidae) at Bukit Kudung in Jeli, Kelantan, Peninsular Malaysia: conservation and ecotourism. Journal of Threatened Taxa 13(1): 17513–17516.

Hedges L, Lam WY, Arceiz AC, Rayan DM, Laurance WF, Latham CJ, Sabaan S and Cle-ments GR (2015) Melanistic leopards reveal their spots: infrared camera traps provide a population density estimate of leopards in Ma-laysia. The Journal of Wildlife Management 79: 846–853.

Jhala YV, Gopal R and Qureshi Q (2018) Status of tigers, copredators and prey in India, 2018 National Tiger Conservation Authority.

Mahabal A, Sharma RM, Patil RN and Jadhav S (2019) Colour aberration in Indian mammals: a review from 1886 to 2017. Journal of Threatened Taxa 11(6): 13690–13719. https: //doi.org/10.11609/jott.3843.11.6.13690-13719

Narayanan J (2020) Assam's 'Golden Tiger' sighting: Not a sign of joy, but concern, says photographer. The Indian Express, July 24, 2020, https: //indianexpress.com/article/lifestyle/life-style/assam-golden-tiger-sighting-mayuresh-hendre-viral-picture-tiger-corridors-eia-6515188/ accessed 24 July 2020.

Palei NC, Rath BP, Palei HS and Mishra AK (2018) Occurrence of melanistic leopard in Odisha, eastern India. Cat news, 68: 7-8.

Panwar HS (1979a) A Note on Tiger Census Technique based on Pugmark Tracings. Indian Forester. Special Issue on International Symposium on Tiger : 70-77.

Panwar HS (1979b) Population dynamics and land tenures of tiger in Kanha National Park: 35-47. In: *International Symposium on tiger, India, February 22-24, 1979. Papers, proceedings and resolutions.* Project Tiger, Government of India, New Delhi.417pp.

Parihar R (2021) Leopard Safaris: Spotting the spotted cats. India Today, 17 September 2021.

Prusty BC and Singh LAK (1996a) Trend of population of tiger and leopard in Similipal Tiger Reserve: a conservation concern. Indian Forester, 122(10): 865-868.

Prusty BC and Singh LAK (1996b) Colour variation in Tiger (*Panthera tigris tigris*). Zoos' Print, XI (9): 20-22.

Prusty BC and Singh LAK (1996c) Melanistic Tiger (*Panthera tigris tigris*) from Similipal Tiger Reserve, Orissa. Tiger Paper, Thailand, XXIII (4), Oct.-Dec.1996: 29-31.

Roychoudhury AK and Acharjyo LN (1983) Origin of white tigers at Nandankanan Biological Park, Orissa. Indian J. Exp. Biol. 21: 350-352.

Sagar SR and Singh LAK (1989) Tiger without stripes. Indian Forester, 115 (4): 277-278.

Sagar SR and Singh LAK (1990) Technique to distinguish tracks of Leopard and Tiger cub. Indian Forester, 116(3): 214-219.

Sagar SR and Singh LAK (1991) Technique to distinguish sex of Tiger (*Panthera tigris*) from pug-marks. Indian Forester 117(1): 24-28.

Sagar SR and Singh LAK (1993) Some inference on the biology of Tiger and Leopard in Similipal Tiger Reserve. Abstract in: International Symposium on the Tiger, Delhi., Ministry of Env. and Forests. Page-12.

Sagar V, *et al.* (2021) High frequency of an otherwise rare phenotype in a small and isolated tiger population. PNAS 2021 Vol. 118 No. 39 e2025273118, https: / doi.org/10.1073/pnas.2025273118, 11pages.

Sayyed A, Takalkar S and Mahabal A (2013) Records of the melanistic leopard, *Panthera pardus* (Linnaeus) from Western Ghats area of Maharashtra and Karnataka, India. Journal of the Bombay Natural History Society 110(2): 151.

Sayyed A and Mahabal A (2015) Second record of melanistic leopard *Panthera pardus* (Linnaeus) from Satara, Maharashtra: a case of roadkill. Zoo's Print 30(5): 29.

Sawarkar VB (1987) Some More on tiger tracks. Cheetal 28 (4): 1-8.

Singh LAK and Swain D (2005) Implications of increase in leopard numbers. Indian Forester, 2005, 131(10), 1383-5.

Singh LAK (1979) Results of survey on the status of mugger crocodile (*Crocodylus palustris* Lesson) in rivers Budhabalanga West Deo in Similipal Tiger reserve, Orissa. Typed Report, Dec. 1979. Similipal Tiger Reserve, Jashipur.

Singh LAK (1995) Sex-identification technique (from pugmark) and sex ratio in Tiger: doubts and clarifications. Indian Forester, 121(10): 885-894.

Singh LAK (1997) Colour variation in Tiger: Preliminary insight into evolution and genetics. Zoos' Print 5 : 9-11.

Singh LAK (1998) *Melanistic and other colour aberrations in Tiger and its implications.* Supported by WWF-India. Facilitated at Similipal Tiger Reserve Forest Department, Govt. of Orissa.

Singh LAK (1999) Born Black: The Melanistic Tiger in India. WWF-India, New Delhi. VIII+66pp.

Singh LAK (2000a). *Tracking Tigers: Guidelines for estimating wild tiger populations using the 'Pugmark Technique'*: Revised Edition. WWF Tiger Conservation Programme.

Singh LAK (2000b) Improved approach to Tiger Counting through Pugmarks. Seminar on 'Wildlife in the wake of the third millennium'. Indian Museum Auditorium, Calcutta, 18 January 2000. Accessible at https: //independent.academia.edu/LalaAswiniKumarSingh.

Singh LAK (2000c) Colour aberration in Tiger: Its Biological and Conservation Implications. Summary of talk at National Seminar 'Tiger Tiger', 4-5 August, 2000, Indian Museum, Calcutta. Accessible at https: //independent.academia.edu/LalaAswiniKumarSingh.

Singh LAK (2014) Current Trends of Wildlife Research (with an overview of research in the past and lessons for future) Page 4-5. Abstract in: Book of Abstracts, UGC Sponsored National Seminar, Current Trends of Animal Science Research in India, 21st and 22nd March 2014, North Orissa University, pp 81.

Singh LAK (2015) A scientometric analysis of the trends of information dissemination on 'true albino' and 'white' mammals. J. Bombay Nat. Hist. Soc. 111(3): 216-220.

Singh LAK (2021) Southward Shifting of Tiger (*Panthera tigris*) Movement Areas in Similipal (1989-2002) indicating approach to Carrying Capacity. Accessible at: https: //independent.academia.edu/LalaAswiniKumarSingh.

Sinha KT (1996) Glimpses of Black Leopard. Cheetal 35(1and2): 19–21.

Srivastava SS and Singh LAK (1997) Future trend of tiger population in Similipal Tiger Reserve. *Indian Forester,* 123(10): 902-909.

Srivastava SS and Singh LAK (1998) Tiger: Melanistic and other colour aberrations (A compilation of Publications from Similipal Tiger Reserve).

TNN (2021). Rajasthan: Rare 'pink' leopard spotted in Ranakpur region in Aravallis hills for first time | India News. Timesnownews.com, accessed 10 November 2021.

van Ingen and van Ingen (1941) Variation in colour of tigers and panthers. *Journal of the Bombay Natural History Society* 42(3): 654.

Wikipedia (2020). Black Tiger. https: //en.wikipedia.org/wiki/Black_tiger. Downloaded on 06 December, 2020.

YouTube (2010). two black panthers filmed in India. Oct 22, 2010, https: //www.youtube.com/watch?v=y0Y4hz_wZik

YouTube (2018a). Kabini as we saw it! Feb 14, 2018, https: //www.youtube.com/watch?v=l9w-l9C8mH4

YouTube (2018b). Melanistic leopard in India: Kabini wildlife Safari (Endangered species), Jun 30, 2018. https: //www.youtube.com/watch?v=wYgpL2AgkIs

Chapter 2

An Overview of Conservation History and Mammalian Diversity of Similipal Biosphere Reserve

***Pratyush P. Mohapatra*[1]*, Cuckoo Mahapatra*[2] *and Sushil Kumar Dutta*[3]**

[1]*Zoological Survey of India, FPS Building, Indian Museum Campus, Kolkata–700016, West Bengal*
[2]*P.G. Department of Zoology, Maharaja Sriram Chandra Bhanja Deo University, Baripada, Mayurbhanj–757003, Odisha*
[3]*Department of Zoology, Assam Don Bosco University, Tapesia garden, Kamarkuchi, Sonapur, Tepesia–782402, Assam*
e-mail: pratyush.m@zsi.gov.in

Introduction

Similipal Biosphere Reserve in Odisha represents a significant biodiversity hotspot within the Chhotanagpur biogeographic province (6B). The Eastern Ghats geographically bounds it to the south and the Lower Gangetic Plains to the east, forming a critical ecological transition zone. The reserve's dominant vegetation type comprises Sal (*Shorea robusta*)-dominated moist deciduous forests, interspersed with riparian zones, perennial streams, and grassland meadows, collectively contributing to its high ecological heterogeneity. Geologically, Similipal is characterized by a stratified structure of metamorphic and sub-metamorphic rock formations arranged in concentric patterns. The metamorphic rock strata primarily include granoid gneiss, true gneiss, and mica schists with interspersed pegmatite intrusions. The sub-metamorphic layers comprise shale, hematite, laterites, limestone, calcareous deposits, phyllites, and micaceous schists. These geological substrates influence soil composition, hydrology, and vegetation patterns, thereby shaping the region's ecological dynamics.

Hydrologically, Similipal is a crucial watershed, hosting several perennial streams, including the Budhabalanga, Palpala, East Deo, Salandi (Nekedanacha), West Deo, Khairi, Bhandan, and Khadkei rivers. These streams play an integral role in sustaining the reserve's diverse ecosystems and support both terrestrial and aquatic biodiversity. The riparian zones, such as the Bhadragoda Valley, Tarinvilla, and Patbil, along with smaller swamps and marshes, provide critical habitats for specialized flora and fauna. The reserve's key grassland meadows, including Devasthali, Bachhurichara, Sapaghar, Matughar, Tiktali, Chahala, and Upper Barakamuda, serve as prime grazing areas for herbivorous species and as hunting grounds for apex predators, forming vital components of the food web. These meadows are ecologically significant for maintaining landscape connectivity and enhancing species dispersal. Similipal's ecological complexity and geomorphological uniqueness render it an ideal site for long-term ecological monitoring and scientific research. The interplay between its geological formations, hydrological systems, and biotic communities provides valuable insights into ecosystem processes, species interactions, and conservation strategies for biodiverse landscapes.

The Early History of Similipal

Similipal, once part of the Mayurbhanj Kingdom, transitioned into Odisha (formerly Orissa) as a district on January 1, 1949, following the integration of the Mayurbhanj state into the Union of India on November 6, 1948. The area holds a rich history of forestry and wildlife management intertwined with human settlement. The Chahala region, known for its century-old Eucalyptus trees planted during colonial forestry operations, serves as a living witness to the transformations in Similipal's landscape. These trees were planted around labour camps established for timber operations and settlements, which began clearing forests in 1906, fostering the gradual formation of villages within the region. During this period, the Maharaja of Mayurbhanj granted 250 acres of tax-free 'Sardari-Jagiri' to Sri Peter Dubraj, a labour contractor, to establish permanent camps for timber operations. This initiative brought labourers, predominantly Kols from Ranchi, to settlements like Garh-Similipal in 1922. At that time, Similipal's dense virgin forests, rich in wildlife, faced significant ecological disruption due to logging activities, expansion of settlements, and indiscriminate hunting. The increased interaction between humans and wildlife led to a rise in tiger attacks on timber workers. To address this, a professional tiger hunter from Ranchi was employed, reportedly killing over 400 tigers within a year using poisoned bow traps—a method documented by the forest department.

Historically, Similipal Reserve Forest covered an area of 539 square miles within a total reserve forest expanse of 1,054.23 square miles by 1907-08. Timber operations, guided by Mr. C.C. Hart's 1898-99 prescriptions, focused on felling Sal trees with a girth exceeding six feet. These operations were carried out in both the Eastern and Western circles of Similipal. The state adhered to the Indian Forest Act of 1927 and implemented its own Forest Manual to regulate forest settlements, surveys, and demarcations, with penalties for violations. Similipal also gained prominence for capturing wild elephants for domestication, particularly for use as war animals. During 1933-34, the Maharajas of Jaipur and Panna visited the Khandabura area to witness an elephant capture operation (kheda). Additionally, the commercial utilization of wildlife, including the capture and export of elephants, tigers, sloth bears, deer, and various bird species, generated significant revenue. Records also document human and livestock casualties from wildlife attacks and instances of illegal poaching of species such as tigers, leopards, sloth bears, and snakes.

The Mayurbhanj Forest Fire Prevention Order was passed in 1936-37 to address forest fires. It implemented strict measures to prevent and manage forest fires, including penalties for negligence and provisions requiring individuals to extinguish or report fires. Despite the pressures of commercial forestry, including supplying railway sleepers, the rulers of Mayurbhanj demonstrated a strong commitment to forest protection. They employed a robust network of forest staff and maintained extensive infrastructure, surpassing that of other princely states and British-administered areas, ensuring the long-term sustainability of Similipal's forest ecosystem.

By the early 70' s the serious decline of tigers had come to notice. The first countrywide tiger census was, therefore, organized based on the Co-operation Tiger Census in 1972 summer. It was conducted in the simple principle of tracing the outlines of the pugmarks of tigers on a framed sheet of thin glass invented by SR Choudhury named the Tiger-tracer. The tiger's paw and toes being soft and pliable, leave individually distinctive pugmarks. Footprints of no two tigers are identical and accurate tracing of pugmarks can hence distinguish individuals separately. The 1972 summer census revealed a total of 1827 adult tigers living in nature in the whole of India. They were admittedly too few, and distressingly still in the decline. Many felt concerned in the national and some international forums. In that concern crystallised the Project Tiger. The project contemplates strengthening of protection and intensification of conservation in the eleven Tiger Reserves located purposefully in different life systems and diverse natural habitats. Each tiger reserve has a core area surrounded by its buffer zone. In Similipal, the core area covers 303 km^2 and its buffer zone 2347 km^2, a total of 2750 km^2. The tiger is rigidly protected and looked after scientifically both in the core and in the buffer, and along with the tiger are necessarily conserved also the plant and animal lives of the ecological systems which harbour and sustain it. Project Tiger started in Similipal in1973 with the following objectives: (i) Elimination of all forms of human exploitation and disturbance from the core and rationalization of such activities in the buffer; (ii) restoration of the ecosystem as close to its natural functioning and (iii) to conduct research about wildlife habitat and wild animals to carefully monitor the changes in flora and fauna over the period of time. The Tiger Project had two demarcated zones namely- Core and Buffer. The management of wildlife of the whole project and protection of the core area remained with the Director of the Tiger Project. The habitat study and preparation of a management plan was done by Late S. R. Chaudhury, I.F.S., the first Field Director of Similipal Tiger Project. The model of an advanced strategy of management was coined for the growth of wildlife and improvement of the ecosystem of the Similipal Hill Reserve. The protective staff of Project Tiger effectively controlled the poaching and improved the habitat by developing pasture grounds, salt licks and water bonds. The V.H.F. and network of telephone connections to staff headquarters all over the project area, to communicate the incident of poaching or illicit cutting could be curbed down the forest to the minimum. The supply of weapons to the staff for self-protection from wildlife and poachers/offenders gives enough moral confidence to combat the offenders and poachers. The roads developed over 360 km out of 947 km by the Odisha State Forest Development Corporation Ltd. facilitated the protection and patrolling in the interior areas of the massif round the year.

Similipal soon became familiar in and outside the country for the tigress reared and brought up by Late Saroj Raj Choudhury, the first Field Director of this Tiger Reserve. The cub presented to him by the Kharias of Similipal was maintained for the study of behaviour, habit and instinct as well as aspects of reproductive biology, senses and inter-specific interactions of the free-living tigress. The conservative eco-conservation strategy augmented by the managers of the tiger project resulted in the increase of the number of the highest predators of the biological pyramid, the Tiger, from 17 in 1973 to 99 in 2002. The increase in the population of Royal Bengal Tigers amply speaks of the revived healthy ecological condition where the tiger and other carnivores along with the herbivores live. Any change in the ecosystem will adversely influence the biodiversity, causing loss or even extinction of the species. To reduce the man-animal conflict, a proposal was mooted to evict and resettle the 62 villagers outside the project area as about 95 per cent of the inhabitants are Adivasis. Most of the plain forests have been encroached for agriculture thus shrinking the wildlife habitat. When the animals stray in to their crop field, they kill those and even the elephants with poisoned arrow or bullet shots. More disappointing is that these villagers being directed or accompanied by Shikaries/poachers kill the male elephant for ivory, the tusk. The tusk hunting racket is operating all over the country. During crop season of October - November, the elephants are shot with poisoned arrows in the temple and under the ears as a result of which is grounded after a month or two and succumb to the injury. Till then the gang members of the arrow hitter keep track of the animal who in right moment extract the tusks and hand over those to the trader to get the balance dues.

The Project Elephant was launched in February 1992 by Government of India for Conservation of elephants, to provide protection to the elephant habitats and to restore the existing elephant corridors and in 2001-02 the Mayurbhanj Elephant Reserve was created. Similipal was declared a Biosphere Reserve in June 1994 with the objectives of conserving the area as a representative ecosystem under the Mahanadian Biographic Region, creating provision for long-term conservation of genetic diversity *in situ,* promoting basic and applied research work and monitoring and dissemination of experience for education and training.

In this work, a brief overview is presented based on literature, forest department records, surveys conducted by researchers and personal observations. An updated list of vertebrate faunal diversity of Similipal Biosphere Reserve is presented based on earlier records and observations by the authors. Taxonomic rectification, updated conservation status as per Wild Life (Protection) Amendment Act, 2022 and IUCN Red List Assessment, along with local status and distribution of the species is provided.

Vertebrate Faunal Diversity of Similipal

A total of 61 species of mammals are recorded from the Reserve by various workers and the systematic checklist of mammals, their distribution and conservation status is given in the Table 2.1.

Table 2.1: Mammals of Similipal and their Conservation Status

Sl. No.	*Common English Name and Scientific Name*	*WPA*	*IUCN*	*Local Status*
1.	Asian Elephant *Elephas maximus* Linnaeus, 1758	Schedule-I	EN	Common
2.	Madras treeshrew *Anathana ellioti* (Waterhouse, 1850)	Schedule-II	LC	Uncommon
3.	Rhesus monkey *Macaca mulatta* (Zimmermann, 1780)	Not listed	LC	Common
4.	Northern plain's langur *Semnopithecus entellus* (Dufresne, 1797)	Schedule-II	LC	Common
5.	Three-striped squirrel *Funambulus palmarum* (Linnaeus, 1766)	Not listed	LC	Common
6.	Five-striped squirrel *Funambulus pennantii* Wroughton, 1905	Not listed	LC	Common
7.	Indian giant squirrel *Ratufa indica* (Erxleben, 1777)	Schedule-I	LC	Common
8.	Giant Indian flying squirrel *Petaurista philippensis* (Elliot, 1839)	Schedule-I	LC	Uncommon
9.	Lesser bandicoot *Bandicota bengalensis* (Gray and Hardwicke, 1833)	Not listed	LC	Common

Sl. No.	Common English Name and Scientific Name	WPA	IUCN	Local Status
10.	Large bandicoot *Bandicota indica* (Bechstein, 1800)	Not listed	LC	Common
11.	Blanford's rat *Madromys blanfordi* (Thomas, 1881)	Not listed	LC	Common
12.	Little Indian field mouse *Mus booduga* (Gray, 1837)	Not listed	LC	Common
13.	House mouse *Mus musculus* Linnaeus, 1758	Not listed	LC	Common
14.	House rat *Rattus rattus* (Linnaeus, 1758)	Not listed	LC	Common
15.	Indian long-tailed tree mouse *Vandeleuria oleracea* (Bennett, 1832)	Not listed	LC	Common
16.	Antelope rat/Indian gerbil *Tatera indica* (Hardwicke, 1807)	Not listed	LC	Common
17.	Indian crested porcupine *Hystrix indica* Kerr, 1792	Schedule-I	LC	Common
18.	Indian hare *Lepus nigricollis* F. Cuvier, 1823	Schedule-II	LC	Common
19.	Gray musk shrew *Suncus murinus* (Linnaeus, 1766)	Not listed	LC	Common
20.	White-toothed pigmy shrew *Suncus etruscus* (Savi, 1822)	Not listed	LC	Uncommon
21.	Short-nosed fruit bat *Cynopterus sphinx* (Vahl, 1797)	Not listed	LC	Common
22.	Indian flying fox *Pteropus giganteus* (Brunnich, 1782)	Schedule-II	LC	Common
23.	Indian fulvous bat *Rousettus leschenaulti* (Desmarest, 1820)	Not listed	LC	Common
24.	Rufous horseshoe bat *Rhinolophus rouxii* Temminck, 1835	Not listed	LC	Common
25.	Sri Lanka leaf-nosed bat *Hipposideros lankadiva* Kelaart, 1850	Not listed	LC	Uncommon
26.	Lesser rat-tailed bat *Rhinopoma hardwickii* Gray, 1831	Not listed	LC	Common
27.	Pouch bearing bat *Saccolaimus saccolaimus* (Temminck, 1838)	Not listed	LC	Very rare
28.	Kelaart's pipistrelle *Pipistrellus ceylonicus* (Kelaart, 1852)	Not listed	LC	Common
29.	Indian pipistrelle *Pipistrellus coromandra* (Gray, 1838)	Not listed	LC	Common

Sl. No.	*Common English Name and Scientific Name*	*WPA*	*IUCN*	*Local Status*
30.	Indian pigmy pipistrelle *Pipistrellus tenuis* (Temminck, 1840)	Not listed	LC	Common
31.	Greater yellow bat *Scotophilus heathii* (Horsfield, 1831)	Not listed	LC	Uncommon
32.	Lesser yellow bat *Scotophilus kuhlii* Leach, 1821	Not listed	LC	Uncommon
33.	Indian pangolin *Manis crassicaudata* Gray, 1827	Schedule-I	EN	Rare
34.	Jungle cat *Felis chaus* Schreber, 1777	Schedule-I	LC	Common
35.	Leopard cat *Prionailurus bengalensis* (Kerr, 1792)	Schedule-I	LC	Uncommon
36.	Rusty spotted cat *Prionailurus rubiginosus* (I. Geoffroy S-H, 1831)	Schedule-I	VU	Rare
37.	Fishing cat *Prionailurus viverrinus* (Bennett, 1833)	Schedule-I	EN	Rare
38.	Leopard *Panthera pardus* (Linnaeus, 1758)	Schedule-I	VU	Uncommon
39.	Royal Bengal tiger *Panthera tigris* (Linnaeus, 1758)	Schedule-I	EN	Rare
40.	Common palm civet *Paradoxurus hermaphrodites* (Pallas, 1777)	Schedule-I	LC	Common
41.	Small Indian civet *Viverricula indica* (Desmarest, 1804)	Schedule-I	LC	Common
42.	*Urva edwardsii* (I. Geoffroy S-H, 1818)	Schedule-I	LC	Common
43.	Small Asian mongoose *Urva auropunctatus* (Hodgson, 1836)	Schedule-I	LC	Common
44.	Ruddy mongoose *Urva smithii* Gray, 1837	Schedule-I	LC	Uncommon
45.	Stripe-necked mongoose *Urva vitticollis* Bennet, 1835	Schedule-I	LC	Rare
46.	*Hyaena hyaena* (Linnaeus, 1758)	Schedule-I	NT	Uncommon
47.	Golden jackal *Canis aureus* Linnaeus, 1758	Schedule-I	LC	Common
48.	Gray wolf *Canis lupus* Linnaeus, 1758	Schedule-I	LC	Uncommon
49.	Wild dog/Dhole *Cuon alpinus* (Pallas, 1811)	Schedule-I	EN	Rare

Sl. No.	Common English Name and Scientific Name	WPA	IUCN	Local Status
50.	Bengal fox *Vulpes bengalensis* (Shaw, 1800)	Schedule-I	LC	Uncommon
51.	Sloth bear *Melursus ursinus* (Shaw, 1791)	Schedule-I	VU	Uncommon
52.	Honey badger *Mellivora capensis* (Schreber, 1776)	Schedule-I	LC	Uncommon
53.	Asian small-clawed otter *Aonyx cinereus* (Illiger, 1815)	Schedule-I	VU	Rare
54.	Smooth-coated otter *Lutrogale perspicillata* (I. Geoffroy S-H, 1826)	Schedule-I	VU	Rare
55.	Wild boar *Sus scrofa* Linnaeus, 1758	Schedule-II	LC	Common
56.	Indian mouse deer/Indian chevrotain *Moschiola indica* (Gray, 1852)	Schedule-I	LC	Uncommon
57.	Spotted deer/Chital *Axis axis* (Erxleben, 1777)	Schedule-II	LC	Common
58.	Sambar *Rusa unicolor* Kerr, 1792	Schedule-I	VU	Uncommon
59.	Northern red muntjac/barking deer *Muntiacus vaginalis* (Boddaert, 1785)	Schedule-I	LC	Common
60.	Gaur/Indian bison *Bos gaurus* Smith, 1827	Schedule-I	VU	Uncommon
61.	Four-horn antelope *Tetracerus quadricornis* (Blainville, 1816)	Schedule-I	VU	Uncommon

Order: PROBOSCIDA

Family: ELEPHANTIDAE

The Asian Elephant is one of the charismatic species of Similipal. Earlier reports of Kheda operations in Similipal are testimonials of the commercial use of elephants, narrated by Senapati and Sahu (1967). As per the report, some 200 elephants were captured by Kheda during 1870-71, followed by 40 elephants during 1910, 20 during 1933-34 and 15 in 1937-38. The last capture of elephants from Similipal was by lassoing in 1943. Elephants are found in herds all over Similipal and generally move in moister parts that hold plenty of fodder. The elephants of Similipal represent the East-Central Landscape and their habitats in central India extend over 17,000 km^2 in the states of Odisha, Jharkhand and southern parts of West Bengal. Extensive documentation on the status of Similipal elephants can be found in Swain and Patnaik (2002); Sar and Lahiri Choudhury (2002) and Singh *et al.* (2002). Makhanas are rare in Similipal and an account of Makhnas is detailed by Srivastava *et al.* (1998). The authors also reported a Makana in Masth near the Chahala salt lick on 19.12.1996. Similarly, another was sighted in Bhanjabasa area at Rajabhadi on 04.12.1996. During 1962, a Makhana elephant was also shot by Late Sri Dibakara Bhanja near Rajupal, which was of 8-8.5 in height. In another instance, Swain and Singh (2003) mentioned musth in a female elephant in Similipal.

The elephant distribution landscape of Similipal is connected to Dalma in northern side, Saranda in North-west side, Satkosia forest on the South-west side, Hadagarh Sanctuary in the southern side and Kuldiha sanctuary towards the south-east side. SR Sagar and LAK Singh mentioned that the elephant population of Dalma during 1989 was about 70 in numbers and before 50 years back there was regular migration to Similipal. Growth of Jamsedpur is believed to be the cause of disruption of the migratory route. The Saranda-Similipal migratory route connects through the North Keonjhar-Khiching and there was regular movement of nearly 200 elephants. This route is now not in existence. The Satkosia-Nato R.F.s are contiguous to Similipal and elephants are known to migrate there during monsoon seasons. Connections with Hadagarh and Kuldiha Wildlife Sanctuary with Similipal is well established and now the total area constitutes the Mayurbhanj Elephant Reserve.

There are several migratory routes into and out of Similipal. The well-established routes such as Similipal-Hadgarh-Kuldiha; Similipal-Nato-Sukhuapata-Kuldiha; Tapoban (West Bengal)-Suliapada RF-Shandua RF-Similipal; Musabani (Jharkhand)–Baunsiapat–Joka-Amdiha-Bedhakudar-Bidhubhandara-Similipal; Musabani (Jharkhand)-Gorumahisani-Kaduani-Kanapat-Argalibandha-Bidubhandara-Similipal; Chaibasa-Bhanjabeda RF-Panapataria-Singada-Kadadiha-Similipal; Bhanjabeda RF-Kudarbista-Similipal and Mushaghati (near Satkosia RF)-Kundei-Kathakata (Near Anandapur) are still functional. Of these the Similipal-Hadagarh-Kuldiha is well connected. This corridor covers a total area of 26.2 sq.km and length 16.5 km, width 3.5 m and composed of dry deciduous Sal and miscellaneous forest with pole crops.

Order: SCANDENTIA

Family: TUPAIIDAE

This family comprises one species, Madras tree shrew is reported from Jamuani, Barheipani and Jamuna-Devasthali road, Lulung, Sitakund, Bangriposi and Chahala. This species lives near bushy undergrowth and prefers forest fringe areas. At first sight, it looks like a brown coloured squirrel but escapes quickly to climb a nearby tree.

Order: PRIMATES

Family: CERCOPITHACIDAE

This family is represented by two species: the Hanuman Langur which is moderately distributed in the buffer villages and in the core areas like Upper

Barhakamuda, Nawna, Dhudruchampa and National Park ranges as well. The Rhesus monkey is more frequently encountered in Bangriposi Ghat area.

Order: RODENTIA

Family: SCIURIDAE

Order Rodentia represents Squirrels, Porcupine, Rats and Mice and is represented by 13 species. Family Sciuridae comprises two species of palm squirrels, a species of giant squirrel and the flying squirrel. Indian Palm Squirrel or Five-striped Squirrel is commonly found in the Reserve whereas the Northern Palm Squirrel or Three Striped Squirrel has a patchy distribution, recorded from Kaptipada, Lulung, Bhanjabasa and Chahala. The Giant squirrel is abundantly found in the core ranges of Similipal. Several works on nesting plants, feeding plants and nest material selection have been carried out in much detail (Rout and Singh, 2004, Singh and Rout, 2005). The Malayan Flying squirrel, a nocturnal and cryptic species is rarely encountered in Similipal. During June, 2011 a specimen was found dead at Chahala ring road, probably killed by a raptor. Earlier records of this species as *Petaurista petaurista* was wrongly mentioned in the Similipal checklist (Annonymous, 1999). Family Muridae comprises seven species under six genera. The Lesser Bandicoot and the Large bandicoot are distributed in the agricultural fields as well as in the swampy areas near human habitation, specific records are from Kaptipada, Gudgudia-Barheipani road and Tulasibani-Brundaban road. The Blanford's rat or the White-tailed Wood Rat are seen in the scrub forests and woodlands of the Reserve. The House Rat is commonly seen in villages and human habitations. The Indian Gerbil is recorded from the woodlands of Similipal in UBK, Jamuani, Nawna and Chahala area. Among the mice species reported from Similipal, the Long-tailed Tree Mouse and Little Indian Field Mouse are commonly distributed in the agricultural fields, open fields, human habitation and scrub forests of SBR. Family Hystricidae comprises one species, *i.e.*, Indian crested porcupine which is fairly distributed in the rocky outcrops. Porcupines make their dens in caves and sometimes dig burrows in earthen mounds and under the base of large trees.

Order: LAGOMORPHA

Family: LEPORIDAE

This family is represented by one species, the Indian hare and this species is common in the Reserve. Choudhury (1973) mentioned large scale poaching of the species by the local tribes in SBR. The species is trapped by using wire-noose traps in series as long as 50-100 meters in length.

Order: EULIPOTYPHLA

Family: SORICIDAE

The family is represented by two species; White-toothed pigmy shrew and House shrew. Although the pigmy shrew was not reported earlier from Similipal, this species is not uncommon in the TR. Due to its small size and secretive habit, it might have escaped from the notice of the researchers. This species is diurnal in habit and lives below rock boulders and leaf litter. Specific sighting records of the species are from Lulung, Barhakamuda, Tarinivilla, Chahala and Jashipur. Similarly, the House shrew is distributed in the core as well as buffer areas of the TR in the human habitation. There is also another species of shrew- *Crocidura* sp., which is rarely encountered near Upper Barhakamuda and near Chingudia fall area. The taxonomic identity of the species is yet to be confirmed. However, the genus *Crocidura* has not yet been reported from Odisha.

Order: CHIROPTERA

The order is represented by 12 species under two families (Family: Pteropodidae and Family: Rhinolophidae). Family Pteropodidae comprises three species, namely Short-nosed fruit bat, Indian flying fox and Indian fulvous bat. Family Rhinolophidae is represented by nine species such as Rufous horseshoe bat, Sri Lanka leaf-nosed bat, Lesser rat-tailed bat, Pouch bearing bat, Kelaart's pipistrelle, Indian pipistrelle, Indian pigmy pipistrelle, Greater yellow bat and Lesser yellow bat. Bats are true fliers and they are known to migrate long distances for foraging. The Flying fox and the Indian Pipistril are commonly distributed in the Reserve. Indian Fulvous Fruit Bat, Short-nosed Fruit Bat, Rufous Horseshoe Bat, Lesser Yellow Bat and Indian Pigmy Bat are recorded from Jashipur and Kaptipada areas. The Lesser Mouse-tailed Bat, Greater False Vampire, Greater Yellow House Bat and Black-bearded Tomb Bat are found in the caves.

Order: PHOLIDOTA

Family: MANIDAE

Pangolin is a toothless mammal, have nocturnal activities and lives in burrows. Pangolins are known to be present in several localities in the TR, commonly seen near Chala ring road, Chahala-Barheipani road and Sapaghara. Dens were also found at Lulung, Barhakamuda and Jenabil. The nocturnal behavior, secretiveness and burrowing habit make them a successful animal to hide from their biggest enemy, the humans. Pangolins are poached for their armoured scales, which are believed to have the healing capacity to cure piles and constipation in humans. There are also records of live Pangolin rescue in Baripada division during 1996-97 (one individual) and 2005-06 (one individual).

Order: CARNIVORA

The order is represented by seven families, namely Felidae, Canidae, Hyaenidae, Ursidae, Herpestidae, Mustelidae and Viverridae. Among the Felids, the most charismatic and conspicuous supreme predator, the Royal Bengal Tiger (RBT) is the largest predator in Similipal. This feline is now distributed in some pockets of South Similipal and remarkably abandoned the northern part due to the ever-rising anthropogenic pressure. Tigers are mostly confined to the Upper Barhakamuda,

Jenabil and National Park ranges of Similipal Tiger Reserve and there are occasional reports from the neighbouring ranges like Dhudruchampa and Chahala ranges. Similipal is also known for the "Black Tigers" or the pseudomelanistic tigers. Singh (1993) has given a detailed investigation of the Melanistic tiger and its conservation implications. Leopards are the major competitors of tigers in Similipal. In the Tiger Reserve, leopards are more confined towards the buffer areas and are rarely seen in the core zone inhabited by tigers. Malanistic leopard or Black panther is also recorded in Similipal. Although this morph is only a genetic mutation, its frequency of occurrence and role of other ecological factors need to be verified. Leopard poaching is very common and many cases also remain unnoticed. Leopard is mostly killed for its coat and inter-state link with the local poachers can never be ruled out. The most common lesser cat, the Jungle cat is well known for raiding poultry. It is mostly found near the buffer villages of Similipal. One of the endangered small cat, the Fishing cat, is rarely found in the Reserve. There are reports of fishing cat from Haladia and Joranda area of Similipal. Acharjyo (1993) reported one individual was rescued from the Satkosia forest of Anandapur and brought to Nandankanan Biological Park. Similarly, Leopard cats have been sighted mainly at Brundaban road, Talabandha road, Kiajhari, Tarinivilla, Patbil, Sapaghara and Kabataghai area. Further survey of these lesser cats will unravel their exact status in the Tiger Reserve.

Among the canids, the Wild dog has become locally extinct from Similipal. Some forest managers believe that the wild dogs have become locally extinct due to some kind of epidemic or due to shortage of prey base and some believe that they all got killed by poisoning. However, there are some recent sighting records of the species in the adjacent Kuldiha wildlife sanctuary and they may soon be visiting Similipal. Earlier there were records of 104 packs (400-426 individuals) of wild dogs during 1997 census in Similipal and the sudden decrease in population to 4 packs during the 2004 census is a matter of concern (Tiger Conservation Plan, STR, 2009). Das *et al.* (1993) mentioned about only one individual at Chahala during January 1971 survey by the Zoological Survey of India. Historically, this species was distributed in and around Chahala, Upper Barhakamuda-Meghasani, Gudgudia, Joranda, Makabadi, Bahaghara and Ransa areas. Reports of predation by the wild dogs are from Chahala (a spotted deer on 07.02.83 at Chahala; a wild boar on 21.09.1989 at Haladia pahad; a spotted deer on 23.02.1991 at Chahala campus).

Although no intensive study was carried out on the common species like Golden jackal and Bengal fox, they are recorded from the buffer villages. Jackals and Foxes are more common towards the drier regions of Similipal in Bangriposi, Bisoi, Pithabata (territorial), Udala and Thakurminda ranges of SBR. Both the species live in self-made, under-ground dens near agricultural fields. In 2009, a Fox was found dead near Kendumundi, which was killed by a wire-noose trap, which may be laid for other bush animals. Wolves are elusive animals and generally not considered as a major competitor of Tigers. They prefer dry and degraded habitats and mostly live on livestock. There are records of long migration in wolves as per the study conducted elsewhere in India. Distribution of wolf in Similipal was confined to the buffer ranges of South Similipal as per the previous records but now they are records of sighting in the core areas also. Wolves were seen during 2011-2012 in Kiajhari, Edelbeda of Kendumundi Range; in Barheipani area of Manada range and at Dangadiha of Thakurmunda range. The STR staff also sighted them in Nawna (N), Pithabata (WL) and at Nawna (S) ranges of STR core. Wolves are major predator of livestock, and hence their presence can easily be ascertained from livestock lifting incidents happening in the villages. Based on the information collected from Barheipani village, 21 goats were attacked by wolves in Uski and Barheipani village during the March and April months, 2012 (personal observations). The livestock depredation incidents by wolves were higher at Kendumundi and Satkosia ranges of Karanjia forest division, Manada range of Rairangpur division.

The family Herpestidae comprises mongooses and in Similipal three species are distributed, namely the Grey mongoose, Ruddy mongoose, Small Indian mongoose and the Black-naped Mongoose. The latter species is the largest among its congeners and was recorded from Upper Barhakamuda-Meghasani road, Jenabil and some other parts of the core area.

The family Mustelidae is represented by Honey badger and the otters. In Similipal two species of otters are recorded, namely the Small Indian Otter and the Oriental small-clawed otter. The former species mostly feed on fishes and the latter mostly feeds on crabs. Honey badger is reported from Jamuani and Bhanjabasa road, they are also found in Lulung, Chahala, Pithabata, Tinadiha, Bangriposi, Dangadiha and Udala area. Similarly, Small Indian otter is known from Haldia, Karkachia, Joranda ring road, Bhanjabasa, Upper Barhakamuda and Kandadhenu area. It is also found in the water bodies, and rivers and rivulets flowing through Similipal. The Oriental Small-clawed Otter is distributed in most of the hill streams of the Reserve.

Sloth bear is the only member of Ursidae found in Similipal and is more abundantly present in the fringe area than the core. Sighting and signs of this species are abundant towards the Thakurmunda and Kendumundi ranges of Southern Similipal.

Small Indian civet and the Common Palm Civet of the family Viveridae are commonly found in the TR. Both the species live on trees, inside tree holes and are nocturnal in habit, spending most of the time on ground searching for prey and fruits. A team of researchers from Zoological Survey of India described a new species of Civet from Similipal named *Paradoxerus jorandaensis*, which was later synonymised as a subspecies of Common palm civet (*Paradoxerus hermaphrositus nictans*).

Hyaena (Family: Hyaenidae), the nocturnal scavenger, is recorded from the buffer villages of Similipal and the dens are typically located in the rocky hillocks in the fringe areas of Similipal. This species is commonly seen in the Bangriposi and Bisoi ranges. During 2009-2012 more than five Hyaenas were road killed within a stretch of 5km of NH-5 at Bangriposi Ghat road. Although this species is not a competitor of tigers, they serve as a major scavenger after the Vultures became locally extinct from the landscape.

Order: ARTIODACTYLA

The Family Cervidae is represented by four species, namely Northern red muntjac, Spotted deer, Sambar and Indian chevrotain. A study by Singh and Srivastava in 1999 mentions that the Barking deer is fairly distributed in the Reserve with an estimated population of 4000-5000 individuals in the tiger Reserve with 100 per cent distribution factor. The specific localities are mentioned as Chahala,

Brundaban, Joranda, Nawana, Jenabil, Devasthali, UBK, Bhanjabasa and Gudgudia areas of ther Reserve. The recent taxonomic works on barking deer assigned the species name *M. vaginalis* for the Norther distribution limit of the species. Cheetal or the Spotted deer is a species of plains and Singh and Srivastava (1999) had estimated between 2500-3500 individuals with a distribution factor of 70 per cent in the tiger Reserve. The specific sites of distribution of the species are Chahala, Devasthali, Gudgudia, Bhanjabasa and Bakua. The Sambar is the major prey for tigers and is mostly found in the hill country. This species has been reported common (7000-9000 individuals) with a distribution factor of 100 per cent in the tiger Reserve. Specific localities of Sambar distribution are from Gudgudia, Joranda, Jenabil and UBK ranges. In Similipal Tiger Reserve the main breeding season for sambar is the rainy season, *i.e.*, September–October, and the frequency of sighting very young fawns is very high from the last week of March to the end of April. The smallest among the antelopes, the Mouse deer is commonly found in the Reserve. A population estimation of 1500-1800 individuals with a distribution factor of 100 per cent from Jamuani, Kairakacha, Pansia, Kabataghai and Dudhiani area.

The Wild Pig or Wild boar of Family Suidae is quite common in the Reserve and prefers dense to moderately dense forests. Signs of wild boar feeding are seen in most of the stream banks. However, regular poaching has resulted in severe depletion of this species from many areas of Similipal.

The family Bovidae is represented by two species, the Four-horn antelope and Gaur. The former species is mostly confined to the core area of STR in Devasthali and National Park ranges. Choudhury also mentioned about witnessing the dung heaps in Bachurichara area of UBK and had a pet Chausingha named Manika. Manika was accidentally killed by Khairi while playing with her. Based on the census conducted by the STR management, Chausingha has a moderate population with a distribution factor of 30 per cent in the Reserve. Singh and Swain (2003) mentioned about the distribution of Chausingha in Dhudruchampa, Nigirda-Panasia, near Panasia, Western part of Sunpokhari, Tarinivilla, Nuagaon, Mahaveersal, Patabil, Bachurichara valley, Devasthali, Kiajhari and Bhanjabasa areas of Similipal. However, there is also a recent sighting (2011) of the species from Mahaveersal area of National Park range. The Indian Bison or Gaur was once fairly distributed in the Tiger Reserve (Choudhury, 1973) is now restricted to pockets. Population estimation of 850-900 individuals with a distribution factor of 60 per cent was reported in the tiger Reserve from Brundaban, Karkachia road, Devasthali, Meghasani and Bhanjabasa area.

Conservation

Among the 61 species of mammals reported from the biosphere Reserve, 31 species are listed in Schedule-I, 06 species are in Schedule-II and 24 species are not listed as per the Indian Wild Life (Protection) Amendment Act, 2022, which has been enacted from April, 2023. Similarly, as per the IUCN Red List, five species are Endangered, eight species are Vulnerable, one species is Near Threatened and 47 species are in Least Concern category. The conservation efforts are mostly targeted towards large and charismatic species. Similipal has set exemplary conservation practices in terms of developing and designing studies on population monitoring of large carnivores, elephants and crocodile (mugger). Although Similipal has a protection regime for the entire biodiversity of the region, focus on *in situ* protection and conservation of lesser-known fauna needs to be undertaken.

Use of modern tools and techniques towards conservation of biodiversity, and evaluation of genetic diversity of various taxa to be undertaken. There are scopes of many new species in Similipal and most of the lower groups need taxonomic assessment. As mentioned, there are many unexplored or poorly studied groups of fauna in Similipal on which studies can be undertaken on the taxonomy and biology of the species. Intensive studies on economically and ecologically important species should be focused. Long-term studies on biodiversity pertaining to impact of climate change can be undertaken, which will be of immense help to the manage the ecosystem. Towards conservation of fauna, legal and social interventions can be initiated (through awareness) to deter small scale but intensive hunting of small mammals and birds, forest fire activities, unsustainable MFP harvesting *etc.*. in the Reserve. Time to time conservation action plans for threatened fauna to be exercised as best practices.

Conclusion

Similipal supports a high diversity of mammals due to varied habitat diversity and the protection regime. Out of 85 species of terrestrial mammal recorded from the state, Similipal is home to nearly 72 per cent of the species diversity of the state. However, lesser-known groups such as rodents, bats and shrews need to be surveyed in the Biosphere Reserve, which are yet to be discovered from the area. Other than the large mammals, attention needs to be paid towards conservation of the threatened and rare fauna.

REFERENCES

Acharjyo LN and Mishra R (1975) Occurrence of fishing cat (*Felis viverrina*) in Orissa. J. Bombay Nat. Hist. Soc. 72(1) : 195-196

Acharjyo LN (1998) The six small cats of Orissa. ENVIS (Wildlife and Protected Areas). Wildlife Institute of India, Dehradun 1(2): 18-20.

Acharjyo LN (1999) Status of mustelids, viverrids and herpestids of Orissa. ENVIS Bulletin: Wildlife and Protected Areas 2(2): 62–64.

Ali SM, Chattopadhyaya S, Kankane PL and Gayen NC (1988) A new species of palm civet from Orissa, India. Rec. Zool. Surv. India 85: 45-48.

Anonymous (1999) Checklist of animals: Similipal Tiger Reserve and Similipal Biosphere Reserve (based on compilation by S. S. Srivastava and L. A. K. Singh). Similipal Tiger Reserve, Baripada, India.

Anonymous (2013) Wildlife Odisha, Wildlife Organization, Forest and Env. Dept., Govt. of Odisha: 1-48 pp.

Anonymous (2014) Wildlife Odisha, Wildlife Organization, Forest and Env. Dept., Govt. of Odisha: 1-72 pp.

Anonymous (2015) Wildlife Odisha, Wildlife Organization, Forest and Env. Dept., Govt. of Odisha: 1-72 pp.

Anonymous (2016) Wildlife Odisha, Wildlife Organization, Forest and Env. Dept., Govt. of Odisha: 1-72 pp.

Anonymous (2017) Wildlife Odisha, Wildlife Organization, Forest and Env. Dept., Govt. of Odisha: 1-51 pp.

Anonymous (2019) Wildlife Odisha, Wildlife Organization, Forest and Env. Dept., Govt. of Odisha: 1-88 pp.

Behura BK and Guru GB (1969) Wildlife of Orissa, Prakruti-Utkal University Journal - Science. 6(2): 95-126.

Choudhury SR (1970) Let us count our tigers, Cheetal. 12(2): 41-51.

Choudhury SR (1971) With the tiger tracer. Cheetal. 13(2): 19-25.

Choudhury SR (1972) Tiger census in India. Part – I and II. Cheetal 15(1): 67-84.

Choudhury SR (1974) Management Plan of Similipal Tiger Reserve, pp 126.

Das PK and Agarwal VC (1973) New records of mammals from Orissa. Sci. Cult. 39: 429-430.

Das PK, Lal JP and Agawal VC (1993) Mammalia, Zoological Survey of India. State Fauna Series, 1: Fauna of Orissa., Part 4, Zoological Survey of India, Kolkata, pp 143-180.

Debata S and Palita SK (2019) Odisha's first record of a free-tailed bat (Mammalia: Chiroptera: Molossidae): what could it be?, J. of Threatened Taxa 11(8): 14071–14074.

Debata S, Palita SK and Behera S (2016) Bats of Odisha-A Pictorial Hand Book. Odisha Biodiversity Board, Bhubaneswar, Odisha, India. x +86pp.

Mishra CG, Patnaik SK, Sinha SK, Kar SK, Kar CS and Singh LAK (1996) Wildlife wealth of Orissa. Wildlife Wing, Forest Department, Government of Orissa, Bhubaneswar, India.

Mishra RK, Sahu HK, Nayak AK and Palei HS (2016) Ecology of Similipal Biosphere Reserve. Daya Publishing House Pvt. Ltd., pp 183.

Mohapatra PP, Palei HS and Hussain SA (2014) Occurrence of Asian Small-clawed Otter *Aonyx cinereus* (Illiger, 1815) in eastern India. Current Science 107: 367–370.

Nayak AK, Nair MV and Mohapatra PP (2014) Stripe-necked mongoose *Herpestes vitticolis* in Odisha, Eastern India: a biogeographically significant record. Small Carnivore Conservation (IUCN) 51: 71–73.

Palei HS and Debata S (2019) On the Occurrence of Rusty Spotted Cat *Prionailurus rubiginosus* (I. Geoffroy SaintHilaire, 1831) (Mammalia: Carnivora: Felidae) in Odisha, Eastern India. P. Zool. Soc. India 72(1): 102–105. DOI: 10.1007/ s12595-017-0243-x

Palei HS, Sahu HK and Nayak AK (2015) Population density, diurnal activity pattern and food preference of Indian giant squirrel *Ratufa indica* in Similipal Tiger Reserve, eastern India. Mammal Study 40: 257-263; doi: 10.3106/041.040.0406.

Palei HS, Sahu HK and Nayak AK (2016) Ungulates densities and biomass in Similipal Tiger Reserve, Eastern India. National Ac. Science Letters; doi: 10.1007/ s40009-016- 0470-0.

Palei HS, Palei NC, Rath BP and Mishra AK (2019) Records of the globally threatened Rusty-spotted cat in Odisha, India, Nature Conservation Research. Çàïîâåäíàÿ íàóêà, 4(3): 112–116.

Sagar SR and Singh LAK (1990) Technique to distinguish tracks of leopard and tiger cub. Indian Forester 116 (3): 214-219

Sagar SR and Singh LAK 1991) Technique to distinguish sex of tiger (*Panthera tigris*) from pug-marks. Indian Forester 117 (1): 24-28.

Singh LAK and Swain D (2003) The fourh-horned antelope or Chousingha (*Tetraceros quadricornis*) in Similipal. J. Zoos' Print 18(9): 1197-1198.

Singh LAK (1993) Every big cat leaves its unmistakable mark. Down to Earth. June 15, 1993. pp. 46.

Singh LAK (1995) Sex-identification technique and sex ratio in tiger - doubts and clarifications. Indian Forester 121(10): 885-894.

Singh LAK (1997) Colour variation in tiger: Preliminary insight into evolution and genetics. *J.* Zoos' Print 12(5): 9-11.

Singh LAK (1999) To err is leopard. *in* Pugs - Wild Orissa, Bhubaneswar, 33-37pp.

Singh LAK (2000) Improved approach to tiger counting through pugmarks. The Twilight - Journal of Pugmarks, Calcutta (Kolkata) 2(2-3): 14-22.

Srivastava SS and Singh LAK (1997) Future trend of tiger population in Similipal Tiger Reserve. Indian Forester 123 (10): 902-909.

Srivastava SS and Singh LAK (1998) Conservation implications of black/melanistic tigers in Similipal Tiger Reserve. J. Zoos' Print (9): 48-49.

Srivastava SS, Singh LAK and Acharya BP (1998) Sighting of makhna in Similipal. Indian Forester 124 (5): 361-362.

Chapter 3

Behavioural Analysis of Silk Moth: Similipal Biodiversity a Case Study

Aratibala Sahu[1], Prasanta Kumar Kar[2] and Priya Ranjan Debta[3]

[1]*Department of Zoology, Indira Gandhi Mahila Mahavidyalaya Udala, Mayurbhanj, Odisha*
[2]*Basic Seed Multiplication and Training Centre, Central Silk Board, Ministry of Textile, Korba, Chhatisgarh*
[3]*Post Graduate Department of Zoology M.S.C.B University, Baripada, Mayurbhanj, Odisha*

Introduction

In India, sericulture is as old as the Indian civilization. The knowledge of silkmoth and silk was known to Indians years ago. Nanavaty (1965) stated that India cultivated wild silk independent of China from very ancient times. The domestication of silkworm Bombyx mori is thought to have originated in the Himalayan foot-hill (Thangavelu and Joshi, 1983) and that the Aryans first discovered the silkworm in the Sub-Himalayan regions. Now, Indian silk is known all over the world for its quality, quantity and design. Today, silk and silk making are no more a secret. The whole world knows about it. Presently, sericulture is being practiced in dozens of countries world over. Countries like India, Japan and China has international institutions and training centres to teach sericulture. However, with all these, China has not allowed any other country to surpass its first position in the world silk production.

The natural silk is broadly of two types namely (a) plant silk fibre and (b) animal silk fibre. The natural silk of plant origin is obtained from the members of the family Bombacaceae. The animal silk is further of two types such as (i) mulberry and (ii) non-mulberry. Mulberry silkmoth is a domesticated type whereas non-mulberry is universally known as "wild silk" or "vanya silk". The non-mulberry variety is classified into two types, that is, insect and non- insect types. The insect type of silk is obtained from nine different types of insects. These are named as (i) eri (ii) muga (iii) anaphe (iv) cricula (v) fagara (vi) coan (vii) weaver ant silk (viii) green lacewing fly silk and (ix) tasar. The non-insect type of silk is of two types namely (i) mussel silk obtained from molluscs and (ii) spider silk obtained from spiders.

Insect type of silk is again classified as (i) commercial and (ii) non-commercial type. Eri, muga, tasar, anaphe, fagara and coan are commercial types of insect silk whereas the non-commercial type of silk is produced from the weaver ant and green lacewing fly. The non-insect silk obtained from mussel and spider is non-commercial type. Recently the wild silks, *viz.*, tasar (both tropical and temperate or oak tasar), eri and muga have been given a new name "vanya silk" (Anonymous, 2006). Mulberry silk insect is monophagous whereas the non-mulberry silk insects are polyphagous. Sericulture, the cultivation of silk is an agro-industry, the economic end product of which is silk. It is highly viable, employment creating and income generating industry providing quick returns in shorter duration. Undoubtedly, it plays a pivotal role in reducing unemployment and in assisting the poorest to make a decent living. Today, sericulture is the main source of livelihood either directly or indirectly for a large number of people, the majority of whom are from the economically retarded section of the society – the Scheduled Castes (SCs) and Scheduled Tribes (STs). It is such a product that it is exclusively prepared by poorest of the poor but luxuriously used by the rich. Sericulture, thus, plays an important role in transferring wealth from rich, the consumers, to the poor, the rearers. China is now the biggest silk producing country in the world where about 20 million people are engaged in sericulture and India, ranking second, provides employment to about 5.5 million people. Silk, as an agro-based industry, is a natural fibre for the integration of a set of interdependent rural and semi-urban based activities such as cocoon production, cocoon reeling, yarn twisting, weaving and finally trading. While the cultivation of the food plants and raising of cocoons are agricultural in nature, the production of raw and spun silk are cottage industry based. All these activities result in employment at the rate of 13 work-year per hectare.

The Forest Conservation Act 1980 adjudicates the fact that "tasar cultivation in the forest area by the tribals as a means of their livelihood without taking monoculture, Asan and Arjuna plantation shall be treated as forest activities. Therefore, no prior approval of the Central Government under the act is necessary (Section 1.5 (1))".

The world of silk is mainly constituted by two distinct varieties like mulberry and non-mulberry. The non-mulberry silk constitutes Tasar, Muga, Eri, Anaphe, Cricula, Fagara, Mussel, Spider and Coan silk. Among non-mulberry silks, Eri, Muga and Tasar (both tropical and temperate or oak tasar) were called "wild silk" as these silkworms thrive on nature grown host plants. But, this group has now been given a new identity and is collectively called "vanya silk" (Anonymous 2006). Generally, it is thought that silk refers only to mulberry silk whose insects, Bombyx mori, are reared on mulberry leaves and is a high priced valuable and delicate textile fibre. The rearing practices of tasar silkworms have been studied by the following scientists. Benchamin and Jolly (1986) [1] and Ojha *et al.* (1994) [4] reported on principles of silkworm rearing. The integrated package of rearing and the transfer of the technology to the field has been suggested by Mathur *et al.* (1998) [6], Narain *et al.* (2014), Mohan Rao *et al.* (2014), Satapathy (2014) and Prasad *et al.* (2015). Sinha (2000) made a study on the rearing methods, seed and cocoon production. Some significant studies have been made on the indoor rearing and maintenance of some of the Antheraea species by Alam *et al.,* 1991, Naveen and Savanurmath (2000) [10], Ch. Sudhakar Babu and Purushotham Rao (2000) and Siddiqui *et al.* (2006) [11]. Phenological studies in muga silkworm, Antheraea assama in relation to its rearing and grain age behaviour has been made by Sahu *et al.* (2000). The advantages of typical Tiwari grain age tray for better seed production have been discussed by Tiwari (1998). It is suggested that systematic and methodical rearing practice results in better production of tasar silk (Jolly 1972 [2], 1974, 1976, Jolly *et al.,* 1973, 1975, and Prasad *et al.,* 2012, 2015). Sengupta *et al.* (1981) studied the rearing performance of tasar silkmoth in crop III.

The different ecological aspects of tasar silkworms have drawn the attention of the following researchers. Mohanty (1991) and Nayak and Dash (1998) studied the biology, ecology and culture practices of the wild tasar silkmoth *Antheraea paphia* Linn. Danielvskii (1965) [13] proposed the aspect of photo periodism and seasonal development of silk insect. Saxena *et al.* (1997) and Srivastava *et al.* (2000) studied the impact of abiotic factors and microclimates on the qualitative and quantitative characters of A. mylitta. Patro *et al.* (2000) made a study on the influence of pruning frequency on different commercial characters of tasar cocoon in A. mylitta. Sahay and Kapila (1992) commented on the adverse effect of deforestation on tasar culture.

The following scientists have studied the nutritional behaviour of tasar silkworm. Nutritional ecology of the silkworm which helps in the proper selection of the host plants is of fundamental importance so far as consumption, digestion and assimilation of host plant leaves and production of cocoons are concerned (Evan 1939, Hamamura 1959 [12], Hamamura *et al.,* 1962, Ramkishore *et al.,* 2002). Findings on food plants and consumption of leaves in various larvae are recorded by some sericologists. Nayak *et al.* (1985), Mohanty and Nayak (1993), Prasad *et al.* (2004) [5] and Sinha *et al.* (2005) studied the leaf consumption by successive instars of A. mylitta. According to them, food consumtion is directly proportional to age and shows an extraordinary increase during the last stage (5th instar) of larva. Sex-wise consumption and utilization of the leaves of different eco races of tropical tasar silkworm, Antheraea mylitta Drury during fifth instar of first crop is studied by Ojha *et al.* (2000) [3]. The diseases of tasar food plants and their control measures have been stated by Chakrovorty and Das (1979), Patro *et al.* (1998), Alam *et al.* (2004) [5] and Satapathy and Alam (2012). They have suggested the vegetable disinfectant for tasar eggs. Roy *et al.* (2006) reported on the improvement of tasar crop by using leaf surface micro flora to control harmful microbes. However ther is a need of a in depth study to investigate the factor which influence more to the tropical tasar silk moth of Similipal biosphere reserve.

Objectives

- ☆ To study the behavioural traits of Tasar Silk worm in Simlipal Biodiversity.
- ☆ To trace the climatological and geographical factors for the modal eco race.
- ☆ To focus the life cycle of modal eco race in natural habitat.
- ☆ To ascertain habitat for its conservation.

The study site for the observation and analyses was in Similipal Biosphere Reserve and the materials used for this investigation was Modal eco race of tropical tasar silkmoth, Antheraea mylitta Drury. During the course of investigation, the eggs, different stages of larvae from first to fifth instar, pupae and adults of Modal eco race of tropical wild tasar silkmoths Antheraea mylitta were observed. For metrical analysis, the instruments like monopan digital balance, slide callipers, screw gauge and measuring cylinder were used. The room thermometer, wet and dry bulb hygrometer, wall clock and rain gauge were used for recording the environmental parameters like temperature, relative humidity, photoperiod and rainfall respectively at the study site daily. Cocoons of Modal eco race constitute the important material for the detailed morphology, colour, quality and quantity of fibres. Physical characteristics such as length, breadth, weight, shell ratio, filament length, nature and type of loop and peduncle of cocoons were also studied as the material.

Modal is a Shorea based eco race of tropical tasar silkworm Antheraea mylitta Drury, exclusively wild, multiplies sou moto in nature, univoltine, almost disease free, produces the heaviest cocoon with highest silk content among all the sericigenous lepidopterans of world and contributes 19 per cent to raw silk production of India. Modal eco race is characterized by fecundity of eggs laying up to 390 in number; larvae of polyphagous nature but prefer to sal leaves before feeding on other leaves, highly variable in morphological features, clothed with numerous setae on whole body and of varied colour patterns, *viz.*, brownish-yellow, green, leafy green and deep green; cocoons of large, oval upto 7.31 cm in length, 5.38 cm in diameter and 20 cc (male) to 50 cc (female) in volume and of varying colours; peduncle of cocoon strong and stout with one ring; pupae of robust and almost spherical with brunt sienna body colour; cocoon shells of thick and compact, thermo and hygro- proof; higher shell weight (males upto 16.7 gm and females upto 21.5 gm), silk content of 2.375 kg of reeled yarn from 1000 cocoons and silk ratio percentages of 25.68; filament of a single cocoon upto 1600 meters in length, coarse and heavier with average denier of 12; moths either brick red or brown in colour in males and polymorphic in colour, *viz.*, yellow, pink, sepia *etc.*. in females; moths emerge after nine months of diapause period with emergence and coupling percentages of 86 and 42 respectively. There is great variation in the genotypes of Modal in the varied ecological conditions of Similpal forest. there is an urgent need to investigate all the aspects.

Methods

During the study, the climatological parameters such as temperature, humidity, photoperiod and rainfall of the viable eco pockets were taken into account. The temperature was measured by centigrade thermometer. The maximum and minimum temperature was recorded at three different timings such as 6 am, 12 noon and 6 pm every day during the crop period. Then the mean or average value of maximum and minimum temperature was calculated each day. Similarly, the mean value of temperature was recorded for the entire crop period. In this way the maximum and minimum temperature along with its range during the study period were recorded for each crop of three successive years from 2017 June to 2019 June at the study site.

The second climatological factor is relative humidity, which was measured by the wet and dry bulb hygrometer. The dry and wet temperature were measured and recorded with the help of sling psychrometer for measuring wet and dry temperature. The relative humidity was recorded with the help of humidity standard table.

The third regulating climatological parameter for the crop cycle is photoperiod. The length of the day or photoperiod is the time of sunrise and sunset, which was recorded every day in various brood periods. The photoperiod is basically the length of day that is the difference of time from sunset to sunrise. The day length was determined with the help of the clock. On the basis of above method, the photoperiod was recorded daily during brood periods and the average photoperiod was calculated in each crop from 2017 to 2019. The fourth parameter to study the life cycle was rainfall, which was measured by rain gauge at rain gauge station at Gudgudia. The precipitation is stated in units of inches or centimeters that falls per unit of time. The amount of rainfall was recorded by rain gauge at rain gauge station at Gudgudia during the crop period in each day. The average rainfall was calculated and recorded for each crop cycle in successive three years from 2017 to 2019.

For the study of larval growth and development, the tasar cocoons were collected from the study site. The brood period was started from the day of egg laying till the formation of cocoon which indicated the larval span. For analysis of cocoons, the tasar cocoons were collected randomly from the different study site of Similipal Biosphere Reserve.

The grain age performance of Modal eco race was studied in three consecutive years 2017, 2018 and 2019. The cocoons collected from the vicinity of study areas during the month of August were preserved (100 in each place) to evaluate the grain age behaviour and preservation effect. In all the study areas, the cocoons preserved with due care and grain age operations were conducted during September and October. Rearing was conducted at three places, *viz.*, Gudgudia, Sarat and Lulung during rainy seasons on sal and asan plants to study the effect of altitude, season and food plant.

The Similipal forest is divided into three divisions, *viz.*, eastern, southern and north eastern. It is ranging from 50 m to 1000 m ASL and forest is full of sal and other tasar food plant. Due to abundance of available of sal tree the tasar silkworm which are eating sal leaf for food are called Modal Eco-races of A. mylitta D. when reared above 600 meter ASL exhibit univoltine character. Some eco race in ex-situ condition when bring down to level between 400-600 behave bivoltine and called Nalia and when reared within 400 meter ASL behave trivoltine and called Bogai. In all the cases silkworms feed upon sal leaf and called "Modal Gutti" and on the contrary the same species when feed upon asan or arjuna called Daba Gutti in some place.

Results

Behaviour Study

Tasar silk moth *Antheraea mylitta* Drury belongs to the super family Bombycidea and the family Saturniidae. As a holometabolous insect, it exhibits four distinctive stages such as egg, larva, pupa (cocoon) and adult moth in its life cycle. The larva or caterpillar stage is reached when the future flying insect is ready to face the outside world and it bites its way out of the egg in which it was born. After about a week of incubation, tiny larvae come out from eggs. On hatching, these are dull brownish yellow with black head, but after 48 hours the body turns green and brown. After hatching, begins the most precarious journey of its life when the tiny creature must play a deadly game of hide-and-seek to survive. The larvae grow rapidly as they must race against time so as to pupate and fly away before the predators eat them. The total larval periods varies according to the season, altitude and the food plant.

Larval Feeding

The larval stage is the only feeding stage and the larvae are virtually eating machines. They eat not only for their own growth and metamorphosis, but also produce the several meters of silk they require in crafting a supporting platform while moulting. Their silk also serves as an escape device to drop down and away from attack. The most important of all is silk is used to weave the cocoon. Besides this, a female requires additional food reserves for the development of eggs. The larvae of tasar silkmoth are polyphagous feeding mostly on sal (*Shorea robusta*), asan (*Terminalia tomentosa*), arjuna (*Terminalia arjuna*) and sometimes on ber (*Ziziphus jujuba*) and on some other food plants. But the larvae of Modal and Nalia eco races mostly feed on sal. The larvae feed from the margin to the center of leaves, but they never eat the midrib. The younger larvae rest on the leaf margin, and the older ones attach themselves to petioles or even small twigs with the help of their abdominal feet and claspers. The leaf lamina is pulled up between the mandibles by the thoracic legs. The young larvae are generally selective feeders and feed on the young and succulent leaves, which contain low content of crude fibre while the matured larvae ingest mature leaves in which crude fibre content is comparatively high. Every stretch of feeding is followed by a rest periods of about five minutes or of longer duration in winter and unfavourable weather. While resting, the larvae raise head, contract the body and assume a typical sitting posture. Feeding with rest periods continues round the clock, except when the mercury drops below 150C.

Moulting

The tasar larvae are tetra moulters. Their body grows to considerable extent both in size and weight. The skin does not keep pace with the growth of the larva's body. When a new layer of skin is formed beneath the old skin, the larvae stop eating and come to rest. The larvae secure a firm grip on either the underside of leaves (early stages) or the shaded part of twigs (late stages). During the moult, the anterior part of the body remains obliquely suspended, the prothoracic hood is fully stretched and the protruding dark head is bent ventrally inward. The larvae seldom lose their grip, but once detached they cannot reestablish and moult with difficulty. So the larvae are not disturbed during moulting. During ecdysis (moulting), the old skin splits behind the head, and the larvae slowly wriggle free out of the useless vestments. The larva's simple cigar-shaped structure makes this process of skin-shedding or moulting all the more simple. Moulting continues for 23-24 hours depending on the climatic conditions. The freshly moulted larvae have tender and pale skin. It takes a while for the new skin to harden and only after this the larvae resume eating, often by consuming their recently moulted skin.

The stage between two sheddings is known as an instar. Antheraea mylitta larva at rest, and the newly born larvae are the first instars. The rate of feeding is minimal before and after the moult, but it reaches its peak between these two periods.

Behaviour of Moth

The adult stage is reached when the moth emerges from the pupa around late evening and dawn. Thus, it has the advantage of low light or darkness during the most vulnerable stage of its life. The peak period of emergence varies according to season. The newly emerged moth's wings are crumpled, shrunken and limp. Instinctively, it climbs on to the nearest host plant to hang on the underside of a branch, while its wings slowly expand due to the blood being pumped into the hollow wing veins. The limp wings hang freely downward while they hardened. If they are obstructed, the wings get deformed and the moths die because they fail to fly. Once fully expanded, soft wings gradually become stronger. If by now, it is already day, the moths stay hidden. By evening the moths are ready to take off an exciting journey. If it is a male, the moth would set out in search of a newly emerged female whose presence is picked up by the male's supersensitive antennae. This receptor detects specific chemical attractants (pheromones) emitted by the females at amazing distances. Female moths are born loaded with eggs and they remain perched while discharging the chemical attractant to help males locating them to mate.

Copulation normally occurs in nature on the night of emergence. Oviposition starts 2 to 4 hours after depairing and continues both during the day and night till the death of the female moth intervenes. The maximum number of eggs are laid in the first 12 hours of oviposition. The eggs are laid on the underside (dorsal side) of the leaves of preferred food plants. The incubation periods of eggs vary from 7 to 9 days depending on the season.

Moths have no proboscis for these do not feed at all during their short adult life span. All the energy needs are met from their fat reserves stored during the larval stage. Females barely use their wings. Instead, they spend much of their brief life span laying eggs on precisely the same food plants on which they themselves grow up as caterpillars. The males die exactly after mating with the females.

Metrical Analysis of Modal

Modal is Shorea based eco race of tropical tasar silkworm Antheraea mylitta Drury, exclusively wild, multiplies suo moto in nature, univoltine, almost disease free, produces the heaviest cocoon with highest silk content among all the sericogenous lepidopterans of the world and contributes 19 per cent to raw silk production of India. There is great variation in the genotypes of Modal in the varied ecological conditions of Similipal forest.

Modal eco race is characterized by fecundity of eggs laying up to 390 in number, larva of polyphagous nature but prefers to Sal leaves before feeding on other leaves, highly variable in morphological features, clothed with numerous setae on whole body and of varied coloured patterns, *viz.*, brownish yellow, green, leafy green and deep green. The cocoons are large, oval up to 7.31 cm in length, 5.38 cm in diameter and 20 cc (male) to 50 cc (female) in volume with varying colours. Peduncles of cocoon are strong and stout with one ring in each pupae are robust and almost spherical with brunt sienna body colour, cocoon shells are thick and compact, thermo and hygro-proof with higher shell weight (males up to 16.7 gm and females up to 21.5 gm), silk content of 2.375 kg of reeled yarn from 1000 cocoons and silk ratio percentage of 25.68; filament of a single cocoon is upto 1600 meter of length, coarse and heavier with average denier of 12, moths either brick red or brown in colour in males and polymorphic in colour, *viz.*, yellow, pink and sepia *etc.*. in females. Moths emerge after nine months of diapause periods with emergence and coupling percentages of 86 and 42 respectively.

Modal exhibits different voltinism and seasonal polymorphism in an altitude gradient of 50 to 1000 mASL characterized by varied environmental conditions. Its voltinism changes when reared out of its natural habitat and exhibits deterioration in its qualitative characters. It behaves univoltine in higher altitude (above 600 mASL) with lower temperature, higher humidity, fewer hours of sunshine and shorter day length. This univoltine eco race is the true Modal. However, it behaves bivoltine in medium altitude (300 to 600 mASL) with medium temperature, higher humidity and medium hours of sunshine and day length. This bivoltine eco race is called "Nalia" which is also Shorea based and multiplies in rainy and autumn seasons suo moto. Qualitatively Nalia eco race cocoons are inferior to that of true Modal univoltine cocoons and characterized by a slender and long peduncle with more than one ring. Nalia is established as a distinct eco race as it has a distinct life cycle and gives rise to progeny. Further, at lower altitude (below 300 mASL) with occurrence of higher temperature and humidity and longer hours of sunshine and day length, the uni and bivoltine eco races like Modal and Nalia behave trivoltine. These trivoltine cocoons are not grown in nature. These are the F1 generation of Modal and Nalia reared for commercial purpose. The diapause cocoons of Modal and Nalia are collected from their natural habitat, reared in lower elevations and their F1 generation is known as "Bogei". Rearers prefer to rearing Bogei because it is totally disease resistant with lower larval mortality, higher cocoon and silk yield and guaranteed crop and reelers prefer Bogei cocoons for its better reelability, evenness, tenacity and lusture of yarn with denier of 9. However, when the diapause Bogei cocoons are preserved for grain age purpose, erratic emergence percentage during preservation is as high as 77.6 per cent and preservation loss also high up to 46.5 per cent. As such, Bogei cocoons are only used for commercial purpose and not for breeding. For taking up a Bogei crop, each time, the diapause cocoons of Modal and Nalia are collected from nature for seed purpose.

Bionomics of Modal

Bionomics (Gr. bios – life; nomos – law) is the study of the relation of an organism or population of organisms to its environment, animate and inanimate. Tasar silk moth is a unique insect, which exhibits voltanism in its life cycle. Some tasar wild silk moths show one crop, some show two crops and some also show three crops in a crop cycle or year. Basing upon the number of crops or cultivation, the voltine nature is called as univoltine, bivoltine, trivoltine. The life cycle of different insects at different heights (altitudes) seems to vary. The growth, development and life cycle are such phenomena that these are controlled and regulated by environmental factors like temperature, humidity, photoperiods, rainfall and altitude.

Temperature

The first environmental parameter was temperature. The temperature is defined as the degree of hotness or coldness of a body. It can also be defined as that thermal condition of a body, which determines the direction of heat flow when bodies are placed in contact. When two bodies at different temperatures are brought in thermal contact, the flow of heat energy will continue until the temperature of two bodies become equal. The state of equality in the temperature of two bodies is known as thermal equilibrium. In fact, the thermal equilibrium is said to be attained when the heat flowing away from the body is equal to the heat flowing towards the body. Since the Modal eco race is univoltine the maximum and minimum temperature were recorded in the first brood periods only. During the course of investigation, the temperature like maximum, minimum and its highest and lowest temperature were recorded with the help of room thermometer and the average temperature was calculated from 2017 to 2019 in first brood period. In the year 2017 in first brood period during rainy season, the average maximum temperature was found to be 34°C and its range was observed to be 32°C to 36°C. Similarly, the average minimum was noted to be 25°C and that range of minimum temperature was from 25°C to 30°C. In 2018 during above periods, the average maximum temperature was observed to be 32°C whereas the range of maximum temperature was from 30°C to 37°C. The average of minimum temperature was noted to be 27°C whereas that range of minimum temperature was from 22°C to 29°C. In 2009 during above- mentioned periods, the average maximum temperature was found to be 30°C whereas the temperature ranged from 28°C to 32°C. Similarly, the average minimum temperature was recorded to be 23°C whereas the minimum temperature was from 22°C to 25°C. The mean of the average maximum and minimum temperature was calculated to be 32°C and 25°C respectively. The mean temperature during the year 2017 to 2019 in first brood period was observed. The mean of the average maximum temperature was calculated to be 32°C, in first brood period. The mean of the average minimum temperature was calculated to be 25°C in first brood periods from 2017 to 2019.

Table 3.1: Average Temperature of Gudgudia during First Brood Periods (July to August) from 2017 to 2019

Sl.No.	*Year*	*First Brood Periods (July to August)*			
		Maximum Temperature (°C)		*Minimum Temperature (°C)*	
		Average	*Range*	*Average*	*Range*
1	2017	34	32-36	25	25-30
2	2018	32	30-37	27	22-29
3	2019	30	28-32	23	22-25
4	Mean	32	-	25	-

Relative Humidity

Humidity is the state of the atmosphere with respect to its content of water vapour. In fact, warm air is able to hold more water vapour than cold air. When the air holds its maximum amount of water vapour, it is said to be saturated. Absolute humidity is the amount of water vapour in a unit mass of air, usually measured by the grams of water vapour present in one cubic meter of the air (or in grains per cubic foot; 1 grain = 0.0648 gram). The amount of water vapour held by a body of air is governed by temperature and pressure. It becomes saturated at dew point. The lower the temperature, the lower is the absolute humidity. Thus, over the landabsolute humidity is the highest near the equator and the lowest in Central Asia in winter.

Relative humidity (RH) is an index of the amount of water vapour present in the atmosphere. It is the actual vapour pressure expressed as a percentage of the saturation vapour pressure, which would be possible at the same air temperature. Relative humidity is an attempt to measure the readiness with which vapour will condense from the air and is concerned with two variables. The actual water vapour is a given mass of atmosphere and the temperature of that mass of air, which determines the capacity of the air to hold the water vapour. The value of relative humidity varies inversely with temperature and therefore, usually rises during the night, because temperature falls, even though the amount of water vapour may remain constant. It is measured by wet and dry bulb hygrometer. Broadly, relative humidity is the ratio between the actual amount of moisture in the air and that, which would be present if the air were saturated at the same temperature, which is expressed as percentage. Air with an RH of 60 is usually considered as approximately separating dry from moist atmosphere.

It is defined as follows.

$$RH = \frac{\text{Actual quantity of water vapour present in a given volume of air}}{\text{Maximum amount of water vapour in the same volume of air}} \times 100$$

RH is given always in percentage and hence it has no unit. Relative humidity represents the amount of water vapour actually present in air compared with the maximum amount of air held at a given temperature. It tells simply relative content and indicates the degree of saturation of air at a given temperature with water vapour. Relative humidity of saturated air is 100 per cent. Relative humidity was considered as the second climatological factor, which was discussed as follows. The relative humidity was recorded daily by wet and dry bulb hygrometer with the help of relative humidity standard chart during first brood period in rainy season from 2017 to 2019. During rainy season in first brood period, the average relative humidity was observed to be 82 per cent, 80 per cent and 87 per cent in 2007, 2008 and 2009 respectively. But the range was recorded to be 79 to 85 per cent, 78 to 83 per cent and 75 to 80 per cent in 2017, 2018 and 2019 respectively. The mean of the average relative humidity was calculated to be 80 per cent and the range from 2017 to 2019 in first brood periods was from 75 to 85 per cent.

Table 3.2: Average Relative Humidity (RH) of Gudgudia during First Brood Periods (July to August) from 2017 to 2019

Sl.No.	*Year*	*First Brood Periods (July to August) Relative Humidity (per cent)*	
		Average	*Range*
1	2017	82	79-85
2	2018	80	78-83
3	2019	78	75-80
4	Mean	80	75-85

Table 3.3: Meteorological Parameters during Rearing Seasons (September to October)

Sl.No.	*Year*	*Maximum Temperature (°C)*		*Minimum Temperature (°C)*		*Relative Humidity (per cent)*	
		Average	*Range*	*Average*	*Range*	*Average*	*Range*
2	2017	36.0	32-37	25.0	23-28	75	68-93
3	2018	36.6	31-39	23.3	20-27	81	47-92
4	2019	32.0	28-38	21.6	21-29	89	59-100

Photoperiod

The third climatological parameter was photoperiod. The photoperiod is the duration of daylight. When direct solar radiation reaches the earth surface, the relative length of dark and light periods affect the behaviour and growth of animals and plants (Strahler 1975). The length of day is the total hours of daylight perceived during the larval stages were recorded. This included the civil twilight time, that is, the time from sunrise to sun set determined with the help of clock at the study site. During the study periods, the day length was calculated daily as per recording of sunrise and sunset. Then total day length and average day length for a particular brood period were calculated and recorded. In first brood period during rainy season, the total and average day length were recorded to be 384.8hours and 10.04 hours respectively in 2017 but the stated day length were noted to be 372.59hours and 10.07 hours respectively in 2018 whereas these were observed to be 370.37 hours and 10.01 hours respectively in 2019 having the mean value of day length 375.92 hours and 10.04 hours respectively.

Table 3.4: Photoperiod at Gudgudia during First Brood Periods (July to August) from 2017 to 2019

Sl.No.	*Year*	*First Brood Periods (July to August)*	
		Total day Length in Hours	*Average Day Length in hours*
1	2017	384.8	10.04
2	2018	372.59	10.07
3	2019	370.37	10.01
4	Mean	375.92	10.09

Rainfall

The fourth climatological parameter was rainfall. The rainfall was recorded daily with the help of rain gauge at rain gauge station, Gudgudia in first brood period during various seasons from 2017 to 2019. During first brood period in rainy season of 2017, 2018 and 2019, the average rainfall was recorded to be 326mm, 302mm and 375mm respectively.

Table 3.5: Average Rainfall at Gudgudia during First Brood Periods (July to August) from 2017 to 2019

Sl.No.	*Year*	*First Brood Periods (July to August)*
		Rainfall in mm
1	2017	326.0
2	2018	302.0
3	2019	315.0
4	Mean	314.3
5	SD	12.01

From the above-mentioned data, it was observed that the average rainfall during first brood period in rainy season in 2017 was the highest (326.0 mm) in comparison to first brood period in 2018 and 2019. The mean average rainfall during first brood periods in Gudgudia was found to be 314.3 mm.

Metrical Analysis of Modal Cocoon

The cocoons collected from three ecopockets were predominantly blackish grey and yellow in colour with stout peduncle with a clear round ring. The length of the peduncle ranged from 4.57 cm to 9.55 cm, volume 23.62 cc to 40.00 cc, cocoon length 5.05 cm to 5.58 cm, cocoon breadth 3.33 cm to 3.65 cm, cocoon weight 11.20 gm to 17.25 gm, shell weight 2.76 gm to 3.41 gm and SR percentage from 19.59 to 24.64

Table 3.6: Morphometric Parameters of Nature Grown Modal Cocoons

Name of the Eco-pockets	*Sex of Cocoons*	*Volume in cc*	*Length of Cocoons in cm*	*Breadth of Cocoons in cm*	*Peduncle Length in cm*	*Cocoon Weight in gm*	*Shell Weight in gm*	*SR per cent*
Sarat	Female	32.25	5.58	3.59	5.83	16.40	3.41	20.79
	Male	23.62	5.16	3.33	6.17	11.20	2.76	24.64
	Average	27.93	5.37	3.45	6.00	13.80	3.08	22.35
Gudu-gudia	Female	39.80	5.49	3.57	4.57	17.25	3.38	19.59
	Male	33.50	5.16	3.33	6.01	13.42	2.85	21.24
	Average	36.65	5.32	3.45	5.29	15.33	3.11	20.31
Lulung	Female	40.00	5.55	3.65	5.61	17.00	3.35	19.70
	Male	37.00	5.05	3.51	9.55	12.67	3.10	24.45
	Average	38.50	5.30	3.58	7.58	14.83	3.22	21.74

Population Diversity of Eco Race Modal

Five thousand six hundred and fifty-three (5643) nature grown Modal cocoons were collected from three eco pockets during the study periods. The colour polymorphism was observed in the cocoon population. Cocoons are predominantly blackish grey and yellow in colour occurring at a ratio of 82:18 and the male: female ratio was 63:37. Cocoons of different locations did not show any marked variation in cocoon characteristics indicating Modal population to be uniform in characters irrespective of its place of occurrence in the biosphere.

Plant Community Structure and Race-Food Plant Association

Studies on plant community structure were conducted in the identified pockets following quadrant method (10000 sq ft/quadrant). Sal (Shorea robusta) dominated the area (72.70 to 83.95 per cent) followed by Asan (Terminalia tomentosa) (2.3 to 18.7 per cent). 1600 nature grown Modal cocoons collected with twigs of food plants from the forest area of the study sites were studied. Out of which 1552 (97 per cent) cocoons were Sal associated and 48 cocoons (3 per cent) were Asan associated.

Grain Age and Rearing Behaviour of Modal During Second Crop Season at Low Altitude Plain

Modal is mostly univoltine in nature and found in higher altitudes (above 150 m ASL) of Similipal Biosphere Reserve of Orissa. Breeding in nature takes place during March to May followed by a natural rearing. Cocoons in nature are available during July to August. When these cocoons are displaced from nature, these behave as bivoltine in low altitude plains, show moth emergence and breeding. A commercial rearing, thus was taken up in specific areas adjacent to peripheral biosphere zones during September to October resulting in production of cocoons during October to November. These cocoons locally known as "Bogai" are popular among the reelers but do not have any seed value due to continuous and erratic moth emergence during preservation periods. In order to exploit this elite eco-race commercially in other areas of low altitude plains, attempts were made to understand the grain age and rearing behaviour of Modal during second crop season (September to October). The nature grown Modal cocoons were collected from Similipal Biosphere reserve during the month of August of the study periods, *i.e.*, 2017 to 2019 and subjected to indoor grain age at Regional Tasar Research Station (RTRS), Baripada (65 m ASL). Cocoons are sex separation (60:40) were woven into garlands and hung inside nylon net of 6' x 6' x 6' to increase mating. All recommended conditions for indoor grain age were followed during the entire grain age period. A total of three graina ges were conducted in August to September during 2017 to 2019.

Table 3.7: Grain Age Behaviour of Modal at Low Altitude Plains (65 m ASL) Season: August-September

Sl. No.	*Year*	*Cocoons Proce-ssed*	*Moth Emergence*		*Total*	*Couplings Obtained*	*DFLS Produced*	*Cocoon: DFL Ratio*
			Male	*Female*				
1	2017	375	105 (32 per cent)	116 (31 per cent)	211 (59 per cent)	30 (26 per cent)	15	25:1
2	2018	350	126 (36 per cent)	106 (30 per cent)	232 (66 per cent)	37 (35 per cent)	35	10:1
3	2019	680	210 (30.9 per cent)	299 (44 per cent)	509 (74.85 per cent)	77 (25.75 per cent)	75	9:1

The DFLS produced from the above grain ages were reared at Regional Tasar Research Station (RTRS) Farm, Rangamatia (65 m ASL) in the second crop season during 2017 to 2019. Asan plants at 8' x 8' spacing prunned during March and maintained under recommended package of practices were utilized for rearing worms after hatching were brushed on Asan plants in such numbers as to minimize handling. All precautionary measures were taken to restrict predator attack. Meteorological parameters were also recorded during the entire rearing period. A perusal of the data indicated moth emergence ranged from 59 to 74.85 per cent. Emergence of male moths varied from 30.9 per cent to 36 per cent and female moths from 30 to 44 per cent. Coupling performance was found to be low in all the grain ages studied which ranged from 25. 75 to 35 per cent. Though there were no traces of pebrine in any of the samples, few layings were found to be low in fecundity. As the coupling behaviour was not satisfactory, dfl recovery was found to be low, *i.e.*, cocoons: Dfl ratio 9:1 to 25:1. Corresponding to the grain ages, three rearings were also conducted during September to October season. Average fecundity ranged from 178 to 232. Hatching of eggs was found to vary from 67.4 to 71.4 per cent. Average larval weight ranged from 27.0 to 34.3 gm. The larval duration was from 36 to 40 days. Cocoon yield was very low ranging from 2 to 7 cocoons per DFL, with a very low effective rate of rearing of 1.5 to 5.7 per cent. The shell ratio (SR per cent), was however, found to be 10.0 to 14.6 per cent.

Indoor grain age of Modal was not found to be satisfactory at low altitude. The race being endemic to Similipal Biosphere Reserve and enjoying all the conducive climatological conditions, perhaps found it difficult in acclimatizing to low altitude plains where frequent fluctuations in temperature and humidity were experienced. Further, being completely wild in nature, the eco-race was not amenable to human interference and handling as is generally done in indoor grain ages. The above factors in combination might have influenced the grain age performance of the eco race resulting in lower coupling rate and poor dfl recovery. This is further corroborated by an in situ grain age conducted at a peripheral Similipal forest area during the same season in 2009 where 84 per cent moth emergence was recorded with a coupling rate of 58 per cent and cocoon: DFL of 4.7:1.

Similarly, poor rearing performance of the eco race may be attributed to the above factors. The race was found to be highly susceptible to diseases like virosis and bacteriosis due to frequent fluctuation in temperature and humidity resulting in crop failure at times. The deterioration in all the economic characters of the cocoons was observed when reared at low altitude. While nature grown Modal cocoons show an SR per cent of 22 to 28 per cent, cocoons produced at low altitude during second crop season showed a maximum SR of 14.6 per cent. The study, thus, indicated that the eco race Modal was very sensitive to changes in climatological conditions and human interference. Second crop season (September to October) which experienced wide and frequent fluctuations in climatological conditions was not found to be suitable for Modal raering in low altitude areas. More studies are required to understand the reasons for deterioration of the economic characters of Modal cocoons produced in this rearing season at low altitude.

Table 3.8: Rearing Behaviour of Modal at Low Altitude (65 m ASL) during September to October

Sl.No.	*Parameters*	*Periods*		
		2017	*2018*	*2019*
1	No. of Dfls reared	6	35	75
2	Average fecundity	178	214	232
3	Percentage of hatching	67.4	68.5	71.4
4	Average larval weight (gm)	34.3	31.8	27.0
5	Larval duration (days)	38	36	40
6	Cocoons harvested	41	219	182
7	Effective rate of rearing (per cent)	5.7	5.3	1.5

Sl.No.	*Parameters*	*Periods*		
		2017	*2018*	*2019*
8	Yield/Dfl	7	6	2
9	Cocoon weight (gm)	11.5	9.6	12.8
10	Shell weight (gm)	1.68	0.96	1.82
11	SR (per cent)	14.6	10.0	14.2

Development of Race Conservation Technique

Tasar eco race Modal is unique in its behaviour. In nature it is predominately univoltine. The breeding takes place during March to May and a natural rearing is followed. Cocoons are available during July to August. These undergo diapause till next March and exceptionally a lesser percentage of cocoons show moth emergence during August to September. But no evidence is available of a following crop. On the other hand, cocoons displaced from nature showed moth emergence followed by breeding and a commercial crop on Asan and Arjun, locally known as Bogai. These Bogai cocoons, thus, produced do not reverse back and are not recycled.

Modal cocoons in nature are very tough and hard shelled with thick peduncle and a clear round ring. They are bigger in size and show colour polymorphism, blackish grey dominating the population. In nature, Modal is predominantly Sal based. Based on the experimental observations a conservation technique for Modal has been developed which is presented below.

Table 3.9: Time Periods for Seed, Cocoon Preservation and Rearing

1. Seed	
Time of seed collection	February to March from Sal forest after leaf fall. Long distance transportation is avoided.
Area of seed collection	Transitional to buffer zones of Similipal Biosphere Reserve
Cocoon characters	-
Cocoon colour	Blackish grey
Cocoon stiffness	Hard and tough uniform cocoons
Cocoon size minimum	Length: 5.0 cm; Breadth: 3.0 cm
Peduncle length	4.3 to 6.5 cm
Peduncle ring	Clear, round and thick ring, diameter 1.08 to 1.14 cm
Cocoon weight minimum	Female – 13.00 gm, Male - 9.00 gm
SR per cent minimum	Female – 18 per cent, Male – 21 per cent
2. Cocoon preservation cum grain age	
Place of preservation and grain age	In transitional to buffer zones. It can also be done inside a grain age house if available
Method of preservation	In hanging garlands inside a nylon net

Time of grain age	March to May
Oviposition	In earthen cups
Fecundity	Above 230 eggs/dfl
Egg disinfection	No disinfection or surface sterilization is required. Eggs can be washed with light soap to remove meconium and toavoid fungal infection
3. Rearing	
Season of rearing	Preferably May to June. If climatic conditions are congenial, rearing can also be taken up in April. Rearing duringAugust to September may be avoided due to pests, predators and diseases
Place of rearing	in situ rearing preferably in the transitional areas
Food plant	Sal (*Shorea robusta*)
Method of brushing	Worm brushing on Sal trees of medium height in compact patches.
Time of brushing	Early in the morning and late afternoon.
Method of rearing	Natural rearing with minimum interference.

Conclusion

India continues to be the second largest producer of silk in the World. The total raw silk production in the country in 2015-16 are28523 MT. Modal is a unique Eco race found in Similipal Biosphere Reserve of District Mayurbhanj Odisha. It predominantly feeds on sal producing very high quality of cocoons rich in silk content. Besides the fecundity of modal is comparatively more than the ruling Daba. The tribal tasar rearers in and around Similipal Biosphere Reserve, Mayurbhanj, Odisha mainly depend on the wild modal cocoons. Due to higher silk yield and higher cost of cocoon in the market, people collect it from the forest and sale it for profit. However due to certain reason this trend is declining in the natural population of Similipal Bioreserve. Thus there is an urgent need to protect it and to develop a well- organized system to make it sustainable for the rural development at the time of alarming trend to decline. However it is required to trace the appropriate forward and backward linkages with necessary technical backup which must provide a gateway to the future prosperity of this culture.

REFERENCES

Benchamin KV and Jolly MS (1986) Principles of silkworm rearing, Proc. Seminar on "Problems of Sericulture", Ed. S. Mahalingama, Madras, pp 63-108.

Chakravorty R and Das BC (1979) Studies on diseases of tasar food plant and their control measure. Annu. Rept. Central Tasar Research, Ranchi, India, pp 139-141.

Danielvskii AS (1965) Photoperiodism and seasonal development of insects. Oliver and Boyd, London: 1-283.

Dash AK and Nayak BK (1990) Studies on some oviposition behaviour of Godamodal eco race of Indian wild tasar silkmoth *A. paphia* L. (Lepidopters: Saturniidae). Proc. 1st Int. Conf. on wild silkmoth, China, No 118: 5Jolly MS (1972) A new technique of tasar silkworm rearing. Indian Silk 11(2): 5-8.

Hamamura Y (1959) Food selection by silkworm larvae. Nature 183: 1746-1747.

Mathur SK, Singh BM, Sinha AK and Sinha BRRP (1998) Integrated package for rearing of *A. mylitta* D. Indian Silk 37(2): 15-18.

Mathur SK, Singhvi SR and Kushwaha RV (2005) Effect of prolonged high temperature and low humidity on fertility and fecundity in tropical tasar silkworm. Indian Silk 43(12): 18-19.

Mohan Rao K, Satapathy B, Mohapatra SC and Kulshrestha V (2004) Grain age and rearing behaviour of Modal in *ex situ* conditions. Indian Silk 43(2): 13- 14.

Mohanty PK (1998) Tropical Tasar Culture in India. Daya Publishing House, New Delhi, pp 1-153.

Narain R, Rath SS, Prasad BC, Alam MO, Rath PC, Mishra D *et al.* (2004) Integrated package for tasar seed cocoon production and preservation. Indian Silk 42(11): 17-18.

Naveen SP and Savanurmath CJ (2000) Innovations in indoor maintenance of tasar silkmoth, *Antheraea mylitta* Drury. Proc. 3rd Int. Conf. on wild silkmoth, Japan and Central Silk Board, India 5: 61-65.

Ojha NG, Sharan SK, Rai S and Pandey PN (2000) Studies on the sex wise consumption and utilization of the leaves of different food plants in different eco races of the tropical tasar silkworm, Antheraea mylitta Drury during fifth instar of the first crop. Proc. 3rd Int. Conf. on wild silkmoth, Japan and Central Silk Board, India 5: 241-245.

Ojha NG, Sinha SS, Singh MK and Sharan SK (1994) Rearing and cocooning of tropical tasar silkworm, *Antheraea mylitta* in indoor condition. Indian J. Wild Silkmoth and Silk 1(2): 257-260.

Sinha RK, Kulashrestha V, Mishra PK and Thangavelu K (1992) Constraints of tasar industry in India. Indian Silk 3(1): 31

Siddiqui AA, Sharma AK, Chauhan TPS and Khatri RK (2006) Breeding Park for tasar silkworm in Uttaranchal: a feasibility. Indian Silk 45(7): 12-15.

Chapter 4

An Annotated Checklist of Spiders (Arachnida: Araneae) of Similipal Biosphere Reserve, Odisha, India

Rakesh Kumar Mohalik[1], Hemant Kumar Sahu[1] and Pratyush P. Mohapatra[2]

[1]*P.G. Department of Zoology, Maharaja Sriram Chandra Bhanja Deo University, Baripada–757003, Odisha*
[2]*Reptilia Section, Zoological Survey of India, F.P.S. Building, Kolkata–700016, West Bengal*
e-mail: pratyush.m@zsi.gov.in

Introduction

Spiders are a fascinating and diverse group of arachnids which has captured the imagination of humans for centuries. With their complex webs, remarkable hunting strategies, and incredible adaptations, spiders are some of the most intriguing creatures in the animal kingdom. A foundational goal in formulating a biodiversity conservation action plan for any given region is to meticulously inventory its diverse species. Globally, the IUCN has not assessed the conservation status of 99.5 per cent of spider species (Seppälä *et al.*, 2018). While recent research has delved into the diversity and distribution of spiders in Odisha, their representation is notably insufficient compared to other regions in India. This discrepancy is attributed to a pervasive lack of attention and negligence in numerous areas within the state.

Similipal Biosphere Reserve, situated in the northern regions of Odisha, India, represents an invaluable ecological hotspot known for its rich biodiversity and unique habitat characteristics. Among the numerous of species that contribute to this diverse ecosystem, spiders, as essential predators and indicators of environmental health, play a crucial role (Michalko and Entling, 2019). However, despite their ecological significance, the systematic documentation of the spider fauna within this Biosphere Reserve remains limited.

Recognizing the necessity for a comprehensive understanding of the spider diversity in the Similipal Biosphere Reserve, this research paper presents a preliminary checklist of spider species found within this region. Through meticulous field surveys and taxonomic investigations, we aim to shed light on the often overlooked yet vital component of the Biosphere Reserve's ecosystem. By providing a detailed account of the various spider taxa residing in this Biosphere Reserve, we contribute to the broader understanding of the Biosphere Reserve's biodiversity and the conservation efforts necessary for its preservation.

Furthermore, this checklist serves as a foundational resource for future research endeavours, conservation strategies, and environmental management policies targeted at preserving the delicate balance of the Similipal Biosphere Reserve. In India, Caleb and Sankaran (2022) originally identified 1900 species belonging to 489 genera in 60 families. However, Singh and Singh (2021) expanded the list to 2344 species across 596 genera, representing 65 families. In Odisha a total of 267 species of spiders described under 162 genera belonging to 42 families were enlisted that have been recorded from only 21 out of 30 districts of Odisha (Singh, 2021). In recent years, several arachnologists recorded several species of spiders from different regions of Odisha, such as Siliwal and Molur (2008) and Siliwal *et al.* (2008a) recorded several spiders in Odisha; Panda *et al.* (2011) recorded 49 species of spiders from Nandankanan National Park in Khordha district; Mohapatra *et al.* (2014) recorded 41 species of spiders from RIE campus, Bhubaneshwar; Palita (2016) and De and Palita, (2018) recorded 50 species from Koraput district. Choudhury *et al.* (2019) enlisted 96 species of spiders from 8 districts of Odisha. Considering the foundational resource for future research, systematic surveys were carried out to document spider diversity from Similipal Biosphere Reserve of Odisha

Materials and Methods

Spiders were collected from various localities from the outskirts of Similipal Tiger Reserve for morphological identification 2010 to 2023 covering all seasons (monsoon: July to October; winter: November to February; and summer: March to June). The spiders from the Tiger Reserve were identified based on photographs taken during the study period. All possible habitats, *viz.*, dense forest, open forest, wetlands, scrub lands, hill forests, agricultural fields and human settlements were sampled. Spiders were identified by examining the specimens under Olympus SZ10 stereo zoom microscope. For identification, female genitalia were dissected and cleaned in concentrated lactic acid for 15–20 minutes. Identification was done using different publications by Tikader (1980, 1982, 1987), Tikader and Malhotra (1980), Barrion and Litsinger (1995), Song *et al.* (1999), Jocqué and Dippenaar-Schoeman (2006), Gajbe (2008), Sebastian and Peter (2009), and Metzner (2017). In the present checklist, attempts have been made to correct the errors in the scientific names of the spiders following WSC (2023).

Results

During the present study 108 species of spiders belonging to 81 genera and 25 families were reported from Similipal Biosphere Reserve. Here, we would like to make a note that in the present checklist 17 spiders are identified only up to genus level only. Amongst these spiders reported in Odisha, species richness of family Araneidae (29 species.) was found to be highest followed by Pisauridae (23 spp.) and Salticidae (7 spp.). Though the present study was for a short period, therefore, there is a need to carry out systematic surveys in the Biosphere Reserve covering all the seasons to get an optimum estimate of spider diversity.

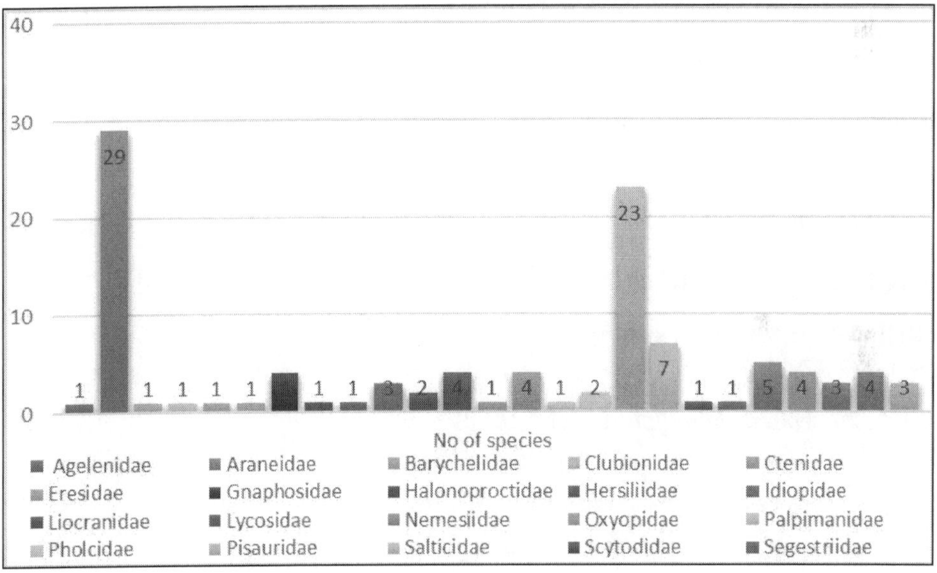

Systematic Checklist of Spiders of Similipal Biosphere Reserve

MYGALOMORPHAE

Family- Barychelidae Simon, 1889

(Trapdoor spiders)

1. *Adelonychia* sp.

Family- Idiopidae Simon, 1889

2. *Idiops nilagiri* Das and Diksha, 2019

3. *Heligmomerus barkudensis* (Gravely, 1921)
4. *Scalidognathus* sp.

Family- Halonoproctidae Pocock, 1901

5. *Conothele varvarti* Siliwal *et al.*, 2009

Family- Nemesiidae Simon, 1892

6. *Gravelyia excavatus* (Gravely, 1921)

Family- Theraphosidae Thorell, 1869

(Tarantula spiders)

7. *Chilobrachys hardwicki* Pocock,1896
8. *Chilobrachys* sp.
9. *Poecilotheria miranda* Pocock, 1900

ARANEOMORPHAE

Family- Agelenidae C. L. Koch,1837

(Funnel web spiders)

10. *Agelena* sp.

Family- Araneidae Clerck, 1757

(Orb web or tent web spiders)

11. *Acusilas* sp.
12. *Araneus bilunifer* Pocock, 1900
13. *Araneus ellipticus* (Tikader and Bal, 1981)
14. *Araneus mitificus* (Simon, 1886)
15. *Araniella cucurbitina* (Clerck, 1757)
16. *Argiope aemula* (Walckenaer, 1837)
17. *Argiope anasuja* Thorell, 1887
18. *Argiope catenulata* (Doleschall, 1869)
19. *Argiope minuta* Karsch, 1879
20. *Argiope pulchella* Thorell, 1881
21. *Bijoaraneus mitificus* (Simon, 1886)
22. *Cyclosa bifida* (Doleschall, 1859)
23. *Cyclosa confraga* (Thorell, 1892)
24. *Cyclosa hexatuberculata* Tikader, 1982

25. *Cyclosa spirifera* Simon, 1889
26. *Cyrtophora cicatrosa* (Stoliczka, 1869)
27. *Eriovixia excelsa* (Simon, 1889)
28. *Eriovixia laglaizei* (Simon, 1877)
29. *Eriovixia poonaensis* (Tikader and Bal, 1981)
30. *Gasteracantha geminata* (Fabricius, 1798)
31. *Gasteracantha kuhli* C.L. Koch, 1837
32. *Gasteracantha unguifera* Simon, 1889
33. *Larinia phthisica* (L. Koch, 1871)
34. *Herennia multipuncta* (Doleschall, 1859)
35. *Neoscona bengalensis* Tikader and Bal, 1981
36. *Neoscona mukerjei* Tikader, 1980
37. *Nephila pilipes* (Fabricius, 1793)
38. *Parawixia dehaani* (Doleschall, 1859)
39. *Poltys nagpurensis* Tikader, 1982

Family- Clubionidae Simon, 1878

(Sac spiders)

40. *Clubiona drassodes* O. Pickard- Cambridge, 1874

Family- Ctenidae Keyserling, 1877

(Wandering spiders)

41. *Ctenus* sp.

Family- Eresidae C. L. Koch, 1845

(Velvet spiders)

42. *Stegodyphus sarasinorum* Karsch, 1891

Family- Gnaphosidae Pocock, 1898

(Ground spiders)

43. *Drassodes luridus* (O.P.-Cambridge, 1874)
44. *Drassodes pashanensis* Tikader and Gajbe, 1977
45. *Gnaphosa pauriensis* Tikader and Gajbe, 1977
46. *Setaphis subtilis* (Simon, 1897)

Family- Hersiliidae Thorell, 1870

(Two-tailed spiders)

47. *Hersilia savignyi* Lucas, 1836

Family- Liocranidae Simon, 1897

48. *Oedignatha microscutata* Reimoser, 1934
49. *Sphingius barkudensis* Gravely, 1931

Family- Lycosidae Sundevall, 1833

(Wolf spiders)

50. *Hippasa agelenoides* (Simon,1884)
51. *Lycosa* sp.
52. *Pardosa* sp.
53. *Wadicosa fidelis* (O.P.-Cambridge, 1872)

Family- Oxyopidae Thorell, 1870

(Lynx spiders)

54. *Oxyope* sp.
55. *Oxyopes shweta* (Tikader, 1970)
56. *Oxyopes sunandae* Tikader 1970
57. *Peucetia viridana* (Stoliczka, 1869)

Family-Palpimanidae Thorell, 1870

(Palp-footed spiders)

58. *Palpimanus* sp.

Family- Pholcidae C. L. Koch, 1850

(Cellar spiders)

59. *Crossopriza lyoni* (Blackwall, 1867)
60. *Pholcus fragillimus* Strand, 1907

Family- Pisauridae Simon, 1890

(Nursery web spiders)

61. *Dendrolycosa gitae* (Tikader, 1970)
62. *Bianor narmadaensis* (Tikader, 1975)
63. *Carrhotus sannio* (Thorell, 1877)

64. *Carrhotus viduus* (*C L Koch, 1846*)
65. *Epeus albus* Proszyñski, 1992
66. *Epeus indicus* Prószyñski, 1992
67. *Habrocestoides bengalensis* Proszyñski, 1992
68. *Hasarius adansoni* (Audouin, 1826)
69. *Hyllus semicupreus* (Simon, 1885)
70. *Menemerus bivittatus* (Dufour, 1831)
71. *Myrmaplata plataleoides* (O.P.- Cambridge, 1869)
72. *Phintella bifurcata* Proszyñski, 1992
73. *Phintella debilis* (Thorell, 1891)
74. *Phintella vittata* (C.L. Koch, 1846)
75. *Phlegra dhakuriensis* (Tikader, 1974)
76. *Plexippus paykulli* (Audouin, 1826)
77. *Plexippus petersi* (Karsch, 1878)
78. *Portia* sp.
79. *Rhene* sp.
80. *Siler semiglaucus* (Simon, 1901)
81. *Siler* sp.
82. *Telamonia dimidiata* (Simon, 1899)
83. *Thiania bhamoensis* Thorell, 1887

Family- Salticidae Blackwall, 1841

(Jumping spiders)

84. *Hasarius adansoni* (Audouin, 1826)
85. *Hyllus semicupreus* (Simon, 1885)
86. *Menemerus bivittatus* (Dufour, 1831)
87. *Phintella vittata* (C. L. Koch, 1846)
88. *Plexippus paykulli* (Audouin, 1826)
89. *Telamonia dimidiata* (Simon, 1899)
90. *Thyene imperialis* (Rossi, 1846)

Family- Scytodidae Blackwall, 1864

(Spitting spiders)

91. *Scytodes pallida* Doleschall, 1859

Family- Segestriidae Latreille, 1804

92. *Indoseges* sp.

Family- Sparassidae Bertkau, 1872

(Giant crab spiders)

93. *Heteropoda hampsoni* Pocock, 1901
94. *Heteropoda phasma* Simon, 1897
95. *Heteropoda venatoria* (Linnaeus 1767)
96. *Olios iranii* (Pocock, 1901)
97. *Olios lamarcki* (Latreille, 1806)

Family- Tetragnathidae Menge, 1866

(Long jawed spiders)

98. *Leucauge decorata* (Blackwall, 1864)
99. *Leucauge fastigata* (Simon, 1877)
100. *Leucauge tessellata* (Thorell, 1887)
101. *Tetragnatha viridorufa* Gravely, 1921

Family- Theridiidae Sundevall, 1833

(Comb-footed spiders)

102. *Achaeraena* sp.
103. *Nihonhimea mundula* (L. Koch, 1872)
104. *Rhomphaea* sp.
105. *Steatoda* sp.

Family- Thomisidae Sundevall, 1833

(Crab spiders)

106. *Camaricus formosus* Thorell, 1887
107. *Oxytate* sp.
108. *Thomisus projectus* Tikader, 1960

Conclusion

This is the first-ever attempt to present a checklist of spiders of Similipal Biosphere Reserve. As per the present report, the spider diversity of Similipal Biosphere Reserve is 40.5 per cent of the known diversity of spiders from Odisha. However, 18 out of the 108 species (16 per cent) were identified only up to the genus level, which may correspond to new species and needs detailed taxonomic investigation. This landscape harbours many interesting species and the diversity comprises representation of Indo-Malayan and Deccan species groups. Systematic surveys of spider fauna will unravel various interesting insights including possibilities of occurrence of endemic spiders from the region.

REFERENCES

Barrion AT and Litsinger JA (1995). *Riceland Spiders of South and Southeast Asia.* CAB International and International Rice Research Institute, Wallingford, England, 736pp.

Caleb JTD and Sankaran PM (2022) Araneae of India, version 2021. https: // indianspiders.in. Retrieved on March 31, 2022

Choudhury SR, Siliwal M and Das SK (2019) Spiders of Odisha: a preliminary checklist. Journal of Threatened Taxa 11(9): 14144-14157. doi: 10.11609/ jott.3786.11.9.14144-14157

De K and Palita SK (2018) A checklist of spiders from six sacred groves in Southern Odisha, India. Serket, The Arachnological Bulletin of the Middle East and North Africa 16(1): 30-40

Gajbe UA (2008) Fauna of India and the Adjacent Countries: Spider (Arachnida: Araneae: Oxyopidae) V. 3. Zoological Survey of India, Kolkata, pp 117.

Jocqué R and Dippenaar-Schoeman AS (2006). Spider Families of the World. Koninklijk Museum Voor Midden-Afrika, pp 336.

Koh JKH and Ming LT (2013) Biodiversity in the Heart of Borneo: Spiders of Brunei Darussalam. Natural History Publications (Borneo), pp 357.

Metzner H (2017) Jumping spiders (Arachnida: Araneae: Salticidae) of the world, online at http: //www.jumping-spiders.com, accessed on 12 July 2017.

Michalko R, Pekár S and Entling MH (2019) An updated perspective on spiders as generalist predators in biological control. Oecologia 189: 21-36.

Mohapatra AK, Biswas T and Parida SP (2014) Spider Diversity in RIE campus. RIE, Bhubaneswar, pp 76.

Palita SK (2016) Faunal diversity assessment of invertebrates and lower vertebrates of Deomali hills of Eastern Ghats, Koraput, Odisha, India. Final Technical Report. Odisha Biodiversity Board and Central University of Orissa, pp 44.

Panda S, Mishra S, Priyadarshini D and Parida P (2011) Spiders of Nandankanan. Forest Department, Government of Odisha, India

Sebastian PA and Peter KV (2009) Spiders of India. Orient Blackswan, Hyderabad, pp 734.

Seppälä S, Henriques S, Draney ML, Foord S, Gibbons AT, Gomez LA, Kariko S, Malumbres-Olarte J, Milne M, Vink CJ and Cardoso P (2018) Species conservation profiles of a random sample of world spiders I: Agelenidae to Filistatidae. Biodiversity Data Journal 6: e23555. doi: 10.3897/BDJ.6.e23555

Siliwal M, Molur S and Daniel B (2008) Status and conservation of tarantulas in selected hot specks of southern India. Wildlife Information Liaison Development (WILD) Society, Coimbatore, Tamil Nadu. Report submitted to Fauna and Flora International, UK

Siliwal M and Molur S (2008) An Inventory of the Spider Fauna of Odisha, India. Unpublished report, submitted to the Odhisa Forest Department, Bhubaneshwar. Zoo Outreach Organisation, India

Singh G and Singh R (2021) Diversity and distribution of crab spiders (Thomisidae: Araneomorphae: Araneae: Arachnida) in India. International Journal of Zoology and Applied Biosciences 6(3): 132-161. https: //doi.org/10.5281/zenodo

Song DX, Zhu MS and Chen J (1999) Spiders of China. Hebei Science and Technology Publishing House, Shijiazhuang, 640pp.

Tikader BK and Malhotra MS (1980) Lycosidae (Wolf-spiders). Fauna India (Araneae) V. 1. Zoological Survey of India, Kolkata, pp 248–447.

Tikader BK (1980) Family Thomisidae. Fauna of India (Araneae) V. 1. Zoological Survey of India, Kolkata, pp 247.

Tikader BK (1982) Family Araneidae (=Argiopidae) and Gnaphosidae. Fauna of India (Araneae) V. 2. Zoological Survey of India, Kolkata, pp 536.

Tikader BK (1987) Handbook of Indian Spiders. Zoological Survey of India, Kolkata, pp 251.

World Spider Catalog (2022) World Spider Catalog. Version 23.0. Natural History Museum Bern. http: //wsc.nmbe.ch. Retrieved on October 23, 2023

Chapter 5

Spatial and Temporal Distribution of Butterflies in the MSCBU Campus, Baripada

***Laxmisagar Dalei*[1], *Sunil Kumar Rout*[1] *and Ranjit Satrusallya*[2]**

[1]*P.G. Department of Zoology, Maharaja Sriram Chandra Bhanja Deo University, Baripada–757003, Odisha*
[2]*P.G. Department of Wildlife and Conservation Biology, Maharaja Sriram Chandra Bhanja Deo University, Baripada–757003, Odisha*
e-mail: ranjitsatrusallya7139@gmail.com

Introduction

The earth exhibits great diversity in the composition of its flora and fauna in different parts of the globe. India is considered one of the 17th mega-biodiversity countries of the world, where the richness of plant as well as animal life is observed. Order Lepidoptera is regarded as one of the important components of biodiversity (New and Collin, 1991). The butterflies constitute the second largest group under the order Lepidoptera in the class Insecta having colorful wing patterns and aesthetic value (Ali and Path, 2000). Further, butterflies are good biological indicators of habitat quality as well as general environment health (Kocher and Williams, 2000; Larsen, 1988; Swachik *et al.,* 2005), as many species are strictly seasonal and prefer only particular sets of habitats (Kunte, 2000). Thus, minor changes in their habitats may lead to either migration or local extinction (Blair, 1999; Kunte, 1997; Mennechez *et al.,* 2003). Butterflies are known to be good pollinators and are ecologically important due to their role in the food chain by acting as both prey and predator for different groups of animals in the forest ecosystem. Their eggs, larvae, and adult stages form the food for various birds and reptiles (Butterfly Ecology, Lawrence E. Gilbert and Michael C. Singer).

India is described as a "butterfly paradise" (Venkataraman, 2010). There are about 19,238 (Heppner, 1998) butterfly species around the world of which India hosts about 1501 species of butterflies. About 200 species of butterflies have been reported from Odisha by different workers. Sethy and Jena (2009) documented 50 species representing 35 genera of 5 families from the Gudgudia range of the Similipal Tiger reserve. Mishra *et al.* (2010) reported 93 species from Nandankanan Wildlife Sanctuary. Nair (2011) recorded 188 species from the Similipal Tiger Reserve. Mohapatra *et al.* (2012) reported 157 species from Bonai. 63 species were recorded from the Utkal University campus by Mohapatra *et al.* (2013). Palei and Rath (2014) reported 101 species from Sunabeda Wildlife Sanctuary. Most recently Payra *et al.* (2016) recorded 136 species belonging to 87 genera from some selected fringe areas of Similipal Biosphere Reserve. Priyamvada and Mohapatra (2016) listed 101 species from the Regional Institute of Education campus, in Bhubaneswar.

Odisha is a junction of four biotic provinces; these are the Eastern and Chhotanagpur plateau under the biogeographic zone Deccan peninsula, the lower Gangetic plain under the zone Gangetic-plain, and the coastline 480km long under the province East coast in the zone of Coasts (Rodgers and Panwar, 1988). Similipal Hills of the state is considered a part of the Eastern Ghats of India (Sinha, 1971). Moreover, Odisha is known as a transitional zone for the flora and fauna of Southern India and Northeast India (Saxena and Brahmam, 1989). This is reflected by the studies on herpetofauna by Dutta *et al.* (2009), and birds and butterflies by Nair (2007). The Maharaja Sriram Chandra Bhanja Deo University (Erstwhile North Orissa University), Takatpur, Baripada, Odisha (21.9303°N 86.7636°E), is located in an urban area having a well-wooded campus amidst a mosaic of concrete buildings. The campus is spread over an area of 98.84 acres with lush green vegetation having large trees, bushy shrubs, and gardens with flowering plants, and long grasses that provide shelter to the butterflies. The good source of nectar, food plants suitable for egg laying, open sunny space, and reduced use of pesticides have resulted in varied species diversity of butterflies on the campus. Such institutional campuses with less pollution and rarely disturbed natural vegetation provide a good site for the preliminary study of the butterfly diversity of a particular area.

The work aims to identify and prepare a checklist of existing butterfly species in Maharaja Sriram Chandra Bhanja Deo University campus and gather information about morphology and the seasonal variation of the documented butterfly species. Though the reported species are confined to a limited number, they can be used as baseline data for further exploration and to assess the impact of habitat destruction caused due to various man-made activities and natural calamities like cyclones that hit the Odisha coast frequently, on butterfly diversity in campus. Accordingly, several conservation strategies can be framed at the local level to maintain the butterfly diversity on campus considering the International Union for Conservation of Nature (IUCN) red list and the Indian Wildlife Protection Act (IWPA) status of the butterfly species.

Methodology

Study Area

Maharaja Sriram Chandra Bhanja Deo University is one of the leading universities of Odisha established on July 13, 1998. It covers a sprawling area of 98.84 acres. Its present campus is located at Takatpur, Baripada in **the** Mayurbhanj district of Odisha. The campus includes administrative and academic buildings, hostels, University staff quarters, temples, small and large ponds, playgrounds, and barren land, connected with **a** good network of roads and flanked by large trees and bushes.

Figure 5.1: Satellite Image of MSCB University Campus.

This survey was done inside the University campus and its surroundings, We were surveyed from July 2022 to May 2023 at several observation sites covering different areas such as Dr. APJ Abdul Kalam Biodiversity Park of the campus, Gardens of Administrative building lawns and gardens and the surrounding areas. The campus of the MSBU is full of vegetation in the planted forest, Small patches of grassland, and also its decorate of a beautiful flower garden. Their surrounding also attracts students for several biodiversitical works. The Ranibandha is out of the university boundary but it attracts many migratory birds and also attracts butterflies and other invertebrates.

The survey was carried out weekly twice during the morning time from 6 am to 9 am, noon time from 12 pm to 1 pm, and afternoon and evening time from 4 pm to 6 pm. The habitats of butterflies in the above-mentioned areas are covered by random observation as well as opportunistic sampling while walking through gardens, lawns, and roads of the University flanked by trees and bushes. When the butterflies were spotted, they were followed till they alighted and came to resting position. Then they were photographed from different angles for proper identification of the specimens. The photographs were taken with Realme 6 pro (64 MP), Honor 9 lite, and Nikon D5600. No specimens were trapped and collected on ethical grounds with a motive to avoid causing harm to them. Identification of butterflies was carried out with the help of field guides (Blyth, 1957; Kunte 2000; Kasambe, 2018).

Results

Previously reported in Similipal Tiger Reserve (STR) 50 species around the 35 genera of 5 families from the Gudugudia range of STR (Sethy and Jena 2009) and Nair in 2011 reported 188 species from Similipal Tiger Reserve. 135 species of butterflies were also reported from Manchabandha and Budhikhamari (Bitupan Bourah *et al.,* 2018). These two planted forests are situated five km—distance from the MSCBU campus and its play a measured role in the path of migratory elephants. In our 11-month survey we found that seasonal variation plays a vital role in the abundance of butterflies and we reported a total of 1318 no. of individuals found in the monsoon, 1006 in winter, and 847 individual in summer. The monsoon attracts more species to the campus and their surroundings. During the study period, 62 species of butterfly were spotted belonging to the five families. Their common names, scientific names, and seasonal occurrences have been compiled in Table 5.1. 25 species belonged to family Nymphalidae (Brush-footed butterflies), 15 species to family Lycaenidae (Blue butterflies), 10 species to family Pieridae (White and yellow butterflies), 6 species to family Hesperidae (Skippers) and 6 species to family Papilionidae (Swallowtails). Seasonal variation study showed that the species richness of butterflies was high during the monsoon season, moderate in the winter season, and less in the summer season (Table 5.2). Out of 62 species of the five families total of 15 species are absent in the summer season, five in the monsoon and 10 in the winter. As for the comparison of the butterfly diversity of the Similipal Biosphere Reserve; Manchabandha and Budhikhamari Reserve Forest, Mayurbhanj, Odisha, India, very less no. of species found in the MSCBU campus but Out of 62 species but 11 species are dose not reported in that areas such as *Melanitis phedima* (Dark evening brown), *Mycalesis subdita* (Tamil Bushbrown), *Rohana parasites* (Black prince), *Discophora sondaica* (Common duffer), *Euploea Sylvester* (Double-branded crow) belongs to the family Nymphalidae; *Tarucus nara* (Striped pierrot), *Tarucus indica* (Indian pierrot), *Spindasis schistacea* (Plumbeous silverline) belongs to the family Lycaenidae; *Oriens goloides* (Small dartlet), *Parnara bada* (Oriental straight swift), *Telicota colon* (Common palm dart) belongs to the family Hesperidae (Table 5.1).

Table 5.1: Checklist of Documented Butterfly Species

Sl. No.	*Family*	*Common Name*	*Scientific Name*	*Season*		
				1	*2*	*3*
1	Nymphalidae	Tawny coster	*Acraea violate* (Linnaeus, 1758)	+	+	+
2		Angled castor	*Ariadne ariadne* (Linnaeus, 1763)	+	+	+
3		Common castor	*Ariadne merione* (Cramer, 1777)	+	+	+
4		Plain tiger	*Danaus chrysippus* (Linnaeus, 1758)	+	+	+
5		Common tiger	*Danaus genutia* (Linnaeus, 1758)	+	+	+
6		Common duffer	*Discophora sondaica* (Boisduval, 1836)	+	+	+
7		Common palm fly	*Elymnias hypermnesia* (Linnaeus, 1763)	+	+	+
8		Common crow	*Euploea core* (Cramer, 1780)	+	+	+
9		King Crow	*Euploea klugii* (Moore, 1857)	-	-	+
10		Common baron	*Euthalia acanthus* (Cramer, 1777)	+	+	+
11		Great egg fly	*Hypolimnas bolina* (Linnaeus, 1758)	+	+	+
12		Peacock pansy	*Junonia almanac* (Linnaeus, 1758)	+	+	+
13		Grey Pansy	*Junonia atlites* (Linnaeus, 1763)	+	+	+
14		Chocolate pansy	*Junonia iphita* (Cramer, 1779)	+	+	+
15		Lemon pansy	*Junonia lemonias* (Linnaeus, 1758)	+	+	+
16		Common evening brown	*Melanitis leda* (Melanitis leda and Linnaeus, 1758)	+	+	+
17		Dark evening brown	*Melanitis phedima* (Cramer, 1780)	-	+	+
18		Commander	*Moduza procris* (Cramer, 1777)	+	+	+
19		Common Bush-brown	*Mycalesis perseus* (Fabricius, 1775)	+	+	+
20		Tamil Bushbrown	*Mycalesis subdita* (Moore, 1890)	-	+	-
21		Long brand bush brown	*Mycalesis visala* (Moore, 1857)	+	+	+
22		Common sailor	*Neptis hylas* (Linnaeus, 1758)	+	+	+
23		Common Leopard	*Phalanta phalantha* (Drury, 1773)	+	+	+
24		Black prince	*Rohana parasites* (Westwood, 1851)	-	+	-
25		Common four-ring	*Ypthima huebneri* (Kirby, 1871)	+	+	+

Sl. No.	Family	Common Name	Scientific Name	Season		
				1	2	3
26	Lycae-nidae	Common pierrot	*Castalius rosimon* (Fabricius, 1775)	+	+	+
27		Lime Blue	*Chilades lajus* (Stoll, 1780)	+	+	+
28		Gram Blue	*Euchrysops cnejus* (Fabricius, 1798)	-	+	+
29		Yamfly	*Loxura atymnus* (Stoll, 1780)	+	+	+
30		Quaker	*Neopithecops zalmore* (Butler, 1870)	+	+	+
31		Pale grass blue	*Pseudozizeeri maha* (Kollar, 1844)	+	+	+
32		Monkey puzzle	*Rathinda amor* (Fabricius, 1775)	+	+	+
33		Plumbeous Silver-line	*Spindasis schistacea* (Moore, 1881)	-	+	+
34		Common silverline	*Spindasis vulcanus (Fabricius, 1775*)	+	+	+
35		Balkan pierrot	*Tarucus balkanicus* (Freyer, 1844)	+	+	+
36		Rounded pierrot	*Tarucus extricatus* (Kollar, 1848)	+	+	+
37		Indian pierrot	*Tarucus indica* (Evans, 1932)	-	+	-
38		Striped pierrot	*Tarucus nara* (Kollar, 1848)	+	+	+
39		Dark grass blue	*Zizeeria karsandra* (Moore, 1865)	+	+	+
40		Tiny grass blue	*Zizula hylax* (Fabricius, 1775)	-	+	+
41	Pieridae	Striped albatross	*Appias libythea* (Fabricius, 1775)	-	-	+
42		Pioneer	*Belenois aurota* (Fabricius, 1793)	-	+	-
43		Common Emigrant	*Catopsilia pomona* (Fabricius, 1775)	+	+	+
44		Molted emigrant	*Catopsilia pyranthe* (Linnaeus, 1758)	+	+	+
45		Common gull	*Cepora Nerissa* (Fabricius, 1775)	+	+	+
46		Jezebel	*Delias eucharis* (Linnaeus, 1758)	-	+	-
47		One spot grass yellow	*Eurema andersonii*	-	+	-
48		Small grass yellow	*Eurema brigitta* (Stoll, 1780)	+	+	+
49		Common grass yellow	*Eurema hecabe* (Linnaeus, 1758)	+	+	+
50		Psyche	*Leptosia nina* (Fabricius, 1793)	+	+	+
51	Hespe-riidae	Common red eye	*Matapa aria* (Moore, 1866)	+	-	+
52		Small dartlet	*Oriens goloides*	-	+	-
53		Oriental straight swift	*Parnara bada*	-	+	-

Sl. No.	*Family*	*Common Name*	*Scientific Name*	*Season*		
				1	*2*	*3*
54	Hespe-riidae	Indian palm bob	*Suastus gremius* (Fabricius, 1798)	+	-	+
55		Dark palm dart	*Telicota bambusae*	-	+	-
56		Common palm dart	*Telicota colon*	+	+	+
57	Papilio-nidae	Tailed Jay	*Graphium Agamemnon* (Linnaeus, 1758)	+	+	+
58		Common jay	*Graphium doson*	-	+	+
59		Crimson Rose	*Pachliopta hector* (Linnaeus, 1758)	+	+	-
60		Common banded peacock	*Papilio crion* (Fabricius, 1793)	+	+	+
61		Lime butterfly	*Papilio demoleus* (Linnaeus, 1758)	+	+	+
62		Common Mormon	*Papilio polytes* (Linnaeus, 1758)	+	+	+

*Note: "+" present, "-" absent, "S(1)" summer, "S(2)" monsoon, "S(3)" winter.

Table 5.2: Family-wise Composition of Butterflies Recorded during the Study Period

Family	*Nymphalidae*	*Lycaenidae*	*Pieridae*	*Hesperidae*	*Papilionidae*
No. of Species	25	15	10	6	6
No. of Genera	16	10	7	5	3
No. of Individuals	**1, 291**	**968**	**762**	**46**	**104**

According to International Union for Conservation of Nature (IUCN) red list data, out of the 62 documented species of butterfly on the MSCB University campus, Seven species belong to the Least concern category, *Castalius rosimon* (Common pierrot) belong to the Vulnerable category (V). and *Discophora spondaic* (Common duffer) belongs to Critically Endangered (CR) category. The status of the 56 butterfly species is Not Evaluated by IUCN. Nine of the total documented butterfly species are legally protected in India under the Wild Life (Protection) Amendment Act, 2022. Of these, 3 species are protected under Schedule I (Part IV), 3 are protected under Schedule II (Part II) and 3 are protected under Schedule IV. Dividing according to Family, 4 species of the Nymphalidae family, 2 species of the Lycaenidae family, 2 species of the Pieridae family, and 1 species of the Papilionidae family come under the Wildlife (Protection) Act, 1972. The highest level of legal protection at the National level is given to them, so these species are very important.

Figure 5.1a

16 17 18 19 20 21 22 23 24 25 26 27 28 29 30 31 32 33 34 35

Figure 5.1b

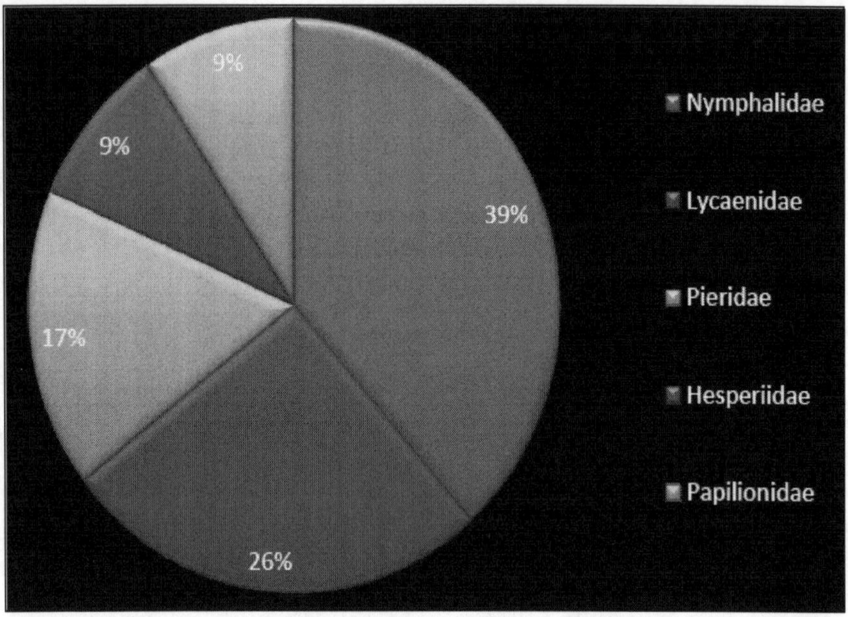

Figure 5.2: Pie Chart Depicting the Percentage of Species in each Family of Documented Butterfly Species in the MSCBU Campus.

The diversified flora at different sites of this campus provided comfortable shelter, suitable foraging grounds, and protection from predators and hostile atmospheric conditions to these butterflies. Further, butterflies feed on the nectar of flowers and their larval stages (caterpillar) depend on different plant leaves to complete their life cycle (Butterfly Ecology, Lawrence E. Gilbert and Michael C. Singer). The availability of a variety of flowering plants, tender buds, leaves, and fruit buds of different plants may fulfill the food requirement of these butterflies. Thus, it is presumed that the diversified flora and fauna of the campus make it rich in butterfly diversity. Hence, the abundance of butterfly diversity in different ecosystems is directly proportional to the type and variety of flowers and the number of plants in a particular area.

As reported by Kunte (2000), an objective revision of the scheduled list is necessary to provide appropriate and adequate legal protection to Indian butterflies. With the increasing needs of the growing human population in India, natural greeneries are being clear-felled giving way to urbanization, pollution, and overgrazing. Loss of prime habitat is a major threat to all wildlife including butterflies. In addition to these, a variety of threats from human recreational activities, trampling, run-off from roads, litter deposition, and weeds are common factors that affect butterfly populations. Although we cannot completely nullify the ill effects of urbanization and development, we can at least try to reduce them by planting endemic trees and plants supporting the local wildlife. This will make sure that at least the common species will not go to the verge of extinction.

Discussion and Conclusion

Apart from the documented 62 species, nine species are under the Indian Wildlife (Protection) Act, 1972 for which utmost care must be taken for their conservation inside the campus which will be a small step to keep these rare species from being pushed to extinction. Some other butterflies were also observed that could not be photographed. Thus, there is all probability that the number of species on the campus is more than reported here. Long-term exploration of butterfly diversity on the campus is thus a necessity to keep a record of the impact of environmental changes on butterfly diversity. Plantation of nectar plants in the gardens and various other sites on a university campus is also suggested to maintain butterfly diversity.

The findings of the present study underline the importance of institutional campuses as a preferred habitat for butterflies. If the landscaping and maintenance of gardens and vegetation inside the campus are carefully planned, the diversity of butterflies may increase in our University campus providing a rich ground for butterfly conservation as well as for research.

REFERENCES

Boruah B, Das GN, Payra A, Gogoi MJ, Dash SK, Tamuly T, Sethy J, Mishra RK, Rout SD (2018) Assessment of Butterfly (Lepidoptera, Rhopalocera) Diversity in Manchabandha and Budhikhamari Reserve Forest, Mayurbhanj, Odisha, India: Asian Journal of Conservation Biology 7(1): 51–65

Biswal AK, Nair MV and Upadhyay HS (2011) Flora and Fauna of Similipal Biosphere Reserve, Floral Diversity: Vol.-1. Similipal Tiger Reserve, Mayurbhanj, Odisha, p 208.

Priyamvada and Mohapatra AK (2016) A preliminary study on the diversity of butterflies (Lepidoptera: Macrolepidoptera) in Regional Institute of Education campus, Bhubaneswar, Odisha, India. Journal of Entomology and Zoology Studies 4(2): 489–496

Basavarajappa S, Gopi Krishna V, Santhosh S (2018) Butterfly species composition and diversity in a protected area of Karnataka, India. International Journal of Biodiversity and Conservation 10(10): 432-443.

Chowdhury S and Das RP (2007). Diversity of Butterflies in the Indian Botanic Garden, Howrah, West Bengal. Bionotes, 9(4): 131-132.

Dasgupta N and Rao RJ (2014) Diversity and seasonal occurrence of butterflies at Jiwaji University campus, Gwalior, Madhya Pradesh. Bugs R All, 2: 16-20.

Dey P, Mondal K and Payra A (2017) A study on butterfly diversity in Singur, West Bengal, India. e-planet, 15 (1): 73-77.

Gilbert LE and Singer MC (1975) Butterfly Ecology. Annual Review of Ecology and Systematics 6 : 365-397

Gupta IJ and Mondal DK (2005) Red Data Book (Part-2): Butterflies of India. Zoological Survey of India, Kolkata.

Harsh S (2014) Butterfly Diversity of Indian Institute of Forest Management, Bhopal, Madhya Pradesh, India. Journal of Insects: 1-4.

Kasambe R (2018) Butterflies of Western Ghats. 2nd Edition. Bombay Natural History Society, India, pp 1-355.

Kawahara AY *et al.* (2019) Phylogenomics reveals the evolutionary timing and pattern of butterflies and moths. PNAS: 1-7.

Kehimkar I (2008) The book of Indian Butterflies, Bombay Natural History Society, Oxford University Press, Walton Street, Oxford, New York.

Kunte K (2000) India – A lifescape: Butterflies of Peninsular India. Indian Academy of Science, University Press, Hyderabad, India, pp 1-254.

Kunte K (2008) The Wildlife (Protection) Act and conservation prioritization of butterflies of the Western Ghats, southwestern India. Current Science 94(6): 729-734

Lamas G (2008) The comparative morphology, phylogeny, and higher classification of butterflies (Lepidoptera: Papilionidea). Kansas Univ. Sci. Bull., 39: 305-370.

Mishra S, Mahapatra PK, Sinha S, Nayak HP, Mishra AK, Nair MV, Mohapatra SN and Panda S (2010). Butterfly diversity of Nandankanan Wildlife Sanctuary, Orissa, India. e-planet 8(1): 20–37.

Mohapatra PP, Sarkar V, Mishra AK and Nair MV (2012) A field guide for beginners: Butterflies of Bonai, Odisha. Odisha Forestry Sector Department Project, Bhubaneswar, Odisha, India.

Mohapatra RK, Mishra AK, Mishra S and Parida SP (2013) A preliminary assessment of Butterfly diversity in Utkal University Campus, Odisha. ZOO's Print, XXVIII (9): 28-31.

Nair AV, Mitra P and Bandyopadhyay SA (2014). Studies on the diversity and abundance of butterfly (Lepidoptera: Rhopalocera) fauna in and around Sarojini Naidu college campus, Kolkata, West Bengal, India. Journal of Entomology and Zoology Studies 2 (4): 129-134.

Nair MV (2007) Butterflies of Similipal Tiger Reserve: A preliminary study on species diversity, species composition and habitat preference. e-planet, 5(1): 76–81.

Padhye AD, Dahanukar N, Paingankar M, Deshpande M and Deshpande D (2006) Season and Landscape distribution of butterflies in Tamhini, Northern, Western Ghats, India. Zoo's Print Journal 21(3): 2175-2181.

Palei NC and Rath BP (2014) Butterflies Diversity of Sunabeda Wildlife Sanctuary, Odisha, India. Journal of Entomology and Zoology Studies 2(2): 39-44.

Perveen F, Khan A and Sikander (2014) Characteristics of butterfly (Lepidoptera) fauna from Kabal, Swat, Pakistan. Journal of Entomology and Zoology Studies 2(1): 56-69.

Sahu HK., Jena J, Dutta SK and Rout SD (2006) Common butterflies of Chahala Range of Similipal Tiger Reserve, Orissa, India. Indian Forester 132(10): 1363–1366.

Umapati Y, Usha DN, Vedavati GN, Girimalleshwar B, Veeranagoudar DK and Pulikeshi MB (2016) Butterfly Diversity of Karnatak University Campus, Dharwad. IOSR Journal of Environmental Science, Toxicology and Food Technology 10 (12): 77-83.

Chapter 6

Traditional Healing Practices among the Hill Kharia Tribe of Similipal Biosphere Region, Odisha

Madhusmita Sahoo

CoE in STMC, Utkal University, Bhubaneswar–751004, Odisha
e-mail: madhusmita.sahoo1@gmail.com

Introduction

Odisha has the distinction of being the homeland of 64 tribal's including 13 Particularly Vulnerable Tribal Groups (PVTGs). In Mayurbhanj district 57.67 per cent tribes constitutes of the total population of this district. There are different tribal groups found in this district and they differ in their skill, aptitudes, habits, cultures, and custom and health systems. Whenever the researcher is going to discuss about the tribe like Hill Kharia, Mankidia, Lodha, Santal, Kolha, Badhudi, Ho *etc.*. Hill Kharia is an important primitive tribe of the Mayurbhanj district, and this is one of the largest indigenous ethnic groups of Odisha. They believe strong faith of the traditional communities on magico-religious action of plants for humane well-being based upon their belief systems. They beliefs in supernatural knowledge or understanding and are mysterious have its roots in *tantra* and they belief in traditional medicine men who prepare medicine from using various types of herbs, horns, teeth, hair, roots and tubers *etc.* available of their local forest, flora and fauna materials for their health. It deals with the health scenario of the Hill Kharia community people observed on their health status that though their mean age at marriage is slightly low, the average fertility of their women is not so high and that the prolonged breast feeding and the prevalence of widespread traditional contraception methods are responsible for their low fertility.

The history of evolution of medicine is a long story stretching from prehistoric to modern times. Many plants used in traditional systems of medicine, usually studied under the purview of ethno medicine, are now recognized to have specific beneficial pharmacological effects. Indeed much of the basic armamentarium of modern pharmacology has been based upon the investigations of the properties of traditional herbal resources. The term 'ethno medicine' is used to refer to those beliefs and practices relating to disease which are the products of indigenous cultural development and are not explicitly derived from the conceptual framework of modern medicine (Chattopadhyay, 2010). Odisha is one of the well-known states in the subcontinent of India having a rich heritage of ethno botanical and ethno medicinal importance.

Ethno-medical practices of any ethnic group are considered important components of their culture, which are transmitted systematically across generations, often through folk means. As a result of several factors including expansion of modern medicine and globalisation, a huge part of ethno-medicine is gradually disappearing thereby endangering part of societies and cultural heritage. Ethno medicine refers to the study of traditional medicine practice. It can encompass methods of diagnosis and treatment. In some cases, it is associated with professional medicine men and women, in others with lay persons who have acquired knowledge from parents or relatives. Herbalism is a traditional medicinal or folk medicine practice based on the use of plants and plants extract. From time immemorial, human diseases and illnesses are said to have originated out of discrepancies related to religious beliefs, magical mysteries and supernatural dogmas and that too having varieties of references in different cultural domain. Illness such as malaria has been conceptualized differently by different communities in traditional societies throughout the globe.

Study Area and the People

The study have been selected for doing research of the Mayurbhanj- a district in the northern Odisha historically famous as the "land of the Maharajas", is also known for its dominant tribal population, vibrant culture, the famous Similpal forest region, Banabasa village of Jasipur Block. In the study area "Banbasa" village where the Hill Kharia primitive tribal people inhabited in the core area of the Similpal forest area and they mostly depend upon the herbal medicine and medicine man for various diseases. General observations on the study area and people provide to understand that the communities rely on their traditional medicines for primary healthcare especially to prevent and cure many kinds of diseases and ailments although the western medicine has reached their doorsteps. The herbal medicines are prescribed by folk or tribal healers, experienced people and also by qualified medical practitioners. Some herbal medicines are prepared out of single ingredient;

part of a plant or whole plant, and some are prepared out of multiple ingredients; parts of plants, parts of animals and minerals.

Mayurbhanj district is the largest tribal communities dwelling of the Santal tribal people and other 16 verities of tribal people inhabiting the same regions. Mayurbhanj district is famous for Similpal biosphere and it is the land of primitive people who are mentioned as indigenous residents of this Indian district include Lodhas, Mankirdias and Kharias. The Mankirdias and Kharias are nomads who spend their time in hunting activities and also food gathering and they reside in the hilly tracts of Similpal Hill Ranges particularly in the sub-division of Panchpir, in the Jashipur blocks.

The Hill Kharia is based on food gathering, hunting and labourer community. The Kharia tribe is divided to sub groups such as – Hill Kharia, Dudh Kharia and Delki Kharia. The Hill Kharia based on their pre agricultural stage of living coming under the PVTG and the Dhelki Kharia is practising agricultural labourers and agriculturalists, while Dudh Kharia are exclusively agriculturists in their primary economy. Hill Kharia people are skilled in cottage industries.

Hill Kharia community still present days pertaining to religious beliefs, magic mysteries and superstition dogmas, that to have a variety of reference in agenesis of their culture before looking into their diseases in respective healing ointment. The villagers were beliefs that the reasons of the diseases that the surrounding environment, seasonal vary and black magic. People mostly prefer to go ***Raulia*** (Magician or ***Gunia***) treatment rather than medicinal practice. People mostly use herb products, local medicine and indigenous medicinal practices for medicinal treatment. Some villagers were unable to express the reason for different diseases and they don't know how to get treatment for a particular disease. In their society the role of women in decision making for health seeking behaviour is very less and most of them are not participating in family decision making. Even in some house they not preferring for medicinal treatment for girls or ladies of that house. It was observed in field that around 61 per cent of households are not allowing women to participation in decision making for health seeking treatment. Tribal culture flourishes in the specific ecological function. According to cultural beliefs and view of the villagers and native health the causes of illness are:

- ✰ Hereditary flow,
- ✰ Religious beliefs,
- ✰ Black magic,
- ✰ Man-made and Seasonal variation.

1. Hereditary Flow

There are few respondents who revel that few diseases like TB, neurological problems and headache are hereditary in nature. They think that some illness occurs due to genetic problem.

2. Religious Beliefs and Cultural Practices of the Herbs

The tribal people have a strong belief and faith on supernatural being may bring any diseases to them if it agrees with their day to day activities. Wrath of the local deities and intrusion of evil spirit is considered as important reasons for various illnesses. As per their perception some religious beliefs like, mythological, supernatural or spiritual aspects of a religion are the causes of illness. The doctrine of karma and re-birth still exists among tribals; they believe that evil dead of parents also make the off spring suffer. If someone indulges in bad deeds, God become angry and inflict some diseases as punishment. There is a common belief found among all the tribal communities that death occurs only to the body of a person, but the soul always remains alive. The soul of their ancestor's roams around their village. If any of them commit any mistake or violate social taboos they used to be punished by their ancestors. The effect of evils eye is considered as one of the important reason for illness. There are some male and female in the village, who poses such evil eye. The role of evil eye in causing disease is very common in study area. It is believed that some individuals cast evil spell on others just by looking at them. Some do it because of jealous of others and desire to possess what other have. Children are believed to be particularly susceptible to the effects of the evil eye.

3. Black Magic

Illness is believed to be caused by human agents through magical means. These categories are divided according to the type of causative agents recognized by tribals. Sorcery is a magical practice and it plays an essential role in the beliefs of the natives as main cause of illness. This act is performed by magicians (***Gunia/ Raulia***) upon some individuals to harm by acquiring body parts like hair, nails *etc.*. The effect of these magical as well as paranormal forces leads a man to become weak, later that resulting in body drying up.

4. Man-made

Tribal's have various notions regarding the food habits as they think that taking more food causes stomach pain. New water in the beginning of rainy season may cause some illness like cold and cough. It is also believed that some kind of pulses high temperature in the body, potato, mulling and sheep meat create cold in the body. Excessive use of local country made liquor alcohol (***handia***) may cause for the bad health and create stomach pain.

5. Seasonal Impact

Certain diseases are found to be caused due to fluctuation of temperature and flow of wind. Prickle is believed to be an epidemic, which occurs by the flow of wind and rise of temperature in rainy season. A disease may be attributed to one cause on one occasion and to another cause on other occasions. Due deep forest and dirty area main cause for many diseases like Malaria, Jaundice, diarrhoea and round warm disease.

6. Folklore and Socio-Religious Beliefs

The tribal people and ethnic races throughout the world have developed their own cultures, custom, cult's religious rites, taboos, totems, legends and myths, folktales and songs, foods medicinal practices *etc.*. Numerous wild and cultivated plants play vital role among these cultures and this relationship has evolved over generations of experience and practices.

The social and economic infrastructure in the villages such as drinking water facilities, electricity, medical facilities, educational institutions, roads, banking and communication networks *etc.* are in very poor and vulnerable condition. Almost all villages are connected with *kuccha* roads *i.e.* link road from residential area in the village to the main road. The problem of drinking water has been a perennial problem in villages. The available well/tube well is insufficient for providing water for drinking, cooking and other uses. Even today, majority of them depend on stream water for drinking and they have to fetch water from long distance.

More than half of the population are illiterate and female literary is in vulnerable state among Hill-Kharia. Educational status beyond primary education is very worse among Hill-Kharia. Food gathering and labour work could not provide regular employment throughout the year. Even today, forest is the main source of survival among Hill-Kharia people. They collect the forest produce for their own consumption and also sell in the market to earn some money. So, forest produce and labour is their main source of income and livelihood.

They don't have not drinking safe water. Their drinking sources are river, pond, stream but these sources are not safe water and they suffer many water diseases like diarrhoea, typhoid, cholera *etc.*. So the Government take step for the safe drinking water for Hill Kharia tribal people, so they dig the hand pump and well in this area.

Table 6.1: Source of Drinking Water of Hill Kharia Tribe on the Study Area

Sl.No.	*Source of Drinking Water*	*No. of Household Per cent (per cent)*
1	Hand Pump	60 (90.91)
2	Stream Water	06(9.09)
3	Grand Total	66 (100)

Table 6.1 explains the availability of drinking water facility in the study area and shown the Hill Kharia tribe 90.91 per cent of the houses depend upon the hand pump water and 9.09 per cent houses depends upon the stream water. The Hill Kharia tribes are enjoying the hand pump and stream water in their live hood.

Common Diseases among the Children

These are some common diseases seen among the Hill Kharia children in the study area. These are discuss below:

Diarrhoeal Disease

Diarrhoea is most common in children, especially those between 6 months to 3 years of age. It is also common in infants under 6 months who are drinking cow's milk or infants feeding formulas. In addition to bringing death, diarrhoea is also a significant contributor to malnourishment in those children who survives. Diarrhoea acts through increased mal-absorption, reduced food intake caused by loss of appetite and food withdrawal, and fever to deprive children of needed nourishment. Diarrhoea is worse and more common in children with malnutrition. Malnutrition is an important element of diarrhoea.

Malaria

Malaria is a major cause of child death in this tribe. Although persons of all ages can get malaria, two groups are at high risk, children under five (0-5) years of age and pregnant women. To reduce the effects of malaria on these high-risk groups, certain types of treatment are necessary. For pregnant women it is important to provide prophylactic treatment with drugs such as chloroquine in order to prevent the risk of abortion of the foetus or low birth weight of neonate. In areas where malaria is endemic, it is recommended that 'presumptive' treatment be given to young a child, which is to treat all young children with drugs when it appears that the fever is due to malaria and that the child will respond to treatment with chloroquine.

Case Study

In the study area Rukumani Dehuri, the age of 35 years old lady residing in Banabasa village with her family. She lost her 5 year old son, due to malaria just 2 years back. He was very healthy. All of sudden he felt severe body pain and was taken to village medicine man. Though, village medicine man had given some medicine in suspense of malaria, he was not cured. The government hospital far from 8 k.m., their home they had not visited due to lack of awareness and illiteracy. After 3 days of severe pain the patient died. Though at last they were advised by the village medicine man to take him to hospital, there was no time.

Malnutrition

Malnutrition was found to be a major health problem in children of Hill Kharia tribe. Malnutrition is used to specifically refer to under nutrition where an individual is not getting enough calories, proteins or micronutrients. The tribal population is at a higher risk of under nutrition because of their dependence on primitive agricultural practices and also transition of occupation as daily wagers and irregularity of food supply. Insufficient food intake leads **Protein Energy Malnutrition (PEM)** and **Chronic Energy Deficiency (CED)**. Low socio-economic status and poor dietary pattern with negligible intake of vegetable and fruits are the main cause of malnutrition among the Hill Kharia children.

The children of the tribal population are particularly vulnerable to malnutrition due to their traditional socio-cultural practices and low literacy level. Education is one of the most powerful means of bringing socio-economic development among the tribe which should be promoted and encouraged.

Case Study

Laxmi Dehuri, at the age group of 24 years old illiterate lady staying in the study village. She lost her son due to malnutrition just 8 months back. One day the boy suddenly became sick. Her mother says that the boy appears 'swollen' and that he was highly irritable. After that he was taken to a *gunia* in their village. *Gunia* claimed that somebody has done some black magic on him and as per his advice they performed some *puja* and sacrificed a goat near village deity. However, all these things failed to save his life. But still she has strong faith on *gunia* and their treatment. Her illiteracy kept her in black to rely on superstition.

Traditional Medicine vs Tribal Medicine

Traditional medicine and tribal medicine vary contextually but they have similarities in many respects. Pal (1986) classified the traditional medicine into three heads:

1. Written traditional systems of medicine (Ayurveda, Unani, Sidha, *etc.*)
2. Oral traditional system of medicine (tribal medicine, family medicine, household remedies, professional and semi-professional medicine)
3. Some components of western medicine

The oral and written traditional systems of medicine are based on transmission of knowledge, whether oral or written transmission. Empirical knowledge of tribals, in general, is transmitted through verbal means and folklore from generation to generation. Like other systems of medicine, tribal medicines have curative and preventive aspects with the application of herbs, animal products and minerals. Their use of medicine is sometimes rudimentary, sometimes very specialized, in certain cases they believe in astrological herbalism and in several cases they take to magic, mysticism, religious performances with or without execution of a sacrifice. The concept of disease and treatment in a tribal society is unique, sometimes simple to understand and sometimes too complex to interpret.

Local Medicine Men and Healers and Therapy in Hill Kharia Tribe

The early healing treatments were derived from the surrounding environment of the human, who were forest dwellers or tribals. They made use of plants, animals and other substances naturally available to them to treat illness. Complex health care system of the simple societies evolved based on deep observation of the nature and environment (Mohapatra, Parida, and Jena). The medical system in simple societies is structured on the lines of herbal and psychometric treatment. The healing practices include a touch of mysticism, supernatural and magic, resulting specific magic-religious rites *etc.*. The health practitioners in Hill Kharia Tribe are ***Raulia*** (medicine man).

Raulia (Witch Doctor)

Raulia is popularly known as shaman. This position is hereditary and it can be achieved by mastering the spell and certain techniques of treatment. As a priest, they worship God (***Adharo deulo Similipal known as deities***) before treatment start. Sometimes they take the patient to the village deity and sacrifice cocks. They also treatment for Malaria, headache and fever.

Table 6.2 shows, there are about 32.44 percent suffer from fever, 18.92 percent people suffer from cold, 21.63 percent suffer from some types of skin disease, 13.51 percent children suffer from malnutrition, 2.70 percent suffer from rheumatism, headache, dental problem, round worm and gastritis problem.

Table 6.2: Showing Different Types of Diseases in Study Area

Sl.No.	*Diseases*	*No. of Patients*	*Per cent*
1.	Cold	07	18.92
2.	Fever	12	32.44
3.	Headache	01	2.70
4.	Rheumatism	01	2.70
5.	Dental problem	01	2.70
6.	Skin disease	08	21.63
7.	Round worm	01	2.70
8.	Gastritis	01	2.70
9.	Malnutrition	05	13.51
10.	Total no. of patients	37	100

Indigenous Medicines for some Diseases Used by Hill Kharia

Hill Kharia has been using flora and fauna since a long time. They gained the knowledge of useful and harmful plants. This knowledge becomes an integral part of their culture. Hill Kharia tribe people are using herbal medicine for their pregnancy related diseases, discuss with Table 6.3.

Conclusion

Traditional medicine is value based, no matter how diverse it is in differential social setting and cultural contexts. There are growth-positive, growth-negative and growth-neutral values in every society associated with traditional healthcare. Such knowledge systems would contribute immensely in the integration and synthesis with any other recognized system of medicine as well help preservation and conservation of a great diversity of important flora. As long as folk uses of the

Table 6.3: Indigenous Medicines for some Diseases of the Study Area

Sl.No.	*Local Name*	*Scientific Name*	*Family*	*Parts Used*	*Diseases*
1.	Amrutabhanda	*Carica papaya*	Caricaceae	Fruit	Lactation
2.	Arakha	*Calotropis gigantia*	Apocynaceae	Root, Latex	Induce abortion and Migration
3.	Ashoka	*Saraca asoca*	Caesalpiniodeae	Bark, Seed, Flower	Irregular menstruation
4.	Ata	*Annona squamosa*	Annonaceae	Root	To enduce abortion
5.	Dalimba	*Punica granatum*	Lythraceae	Root	To stop menstruation
6.	Karanja	*Pongamia pinnata*	Fabaceae	Stem	To control excessive bleeding
7.	Mandara	*Hibiscusrosa-sinensis*	Malvaceae	Stem	For contraception
8.	Satabari	*Asparagus racemosus*	Asparagaceae	Tuber	To initiate lactation
9.	Simili	*Bombax ceiba*	Bombacaceae	Root	To break the sterility of female
10.	Sorisa	*Brassica nigra*	Brassicaceae	Seed	Oil massage in pregnancy

Table 6.4: Distribution of Medicinal Plants Used for Various Ailments by the Hill Kharia Tribe on the Study Area

Local Name	*Scientific Name*	*Family Name*	*Which Parts Used*	*Which Diseases*	*Purification Procedure*	*Preparation Procedure*
Bach	*Acorus Calamus l.*	Araceae	Tuber and root	Tooth pain	Grind with water	Sewing the teeth
Kukura dantadi	*Achyranth esasper l.*	Amarantaceae	Roots	Tooth pain	Grind with water	60 ml applied for leaves taken orally
Bela	*Aeglemar melos (l) corr.*	Rutaceae	seeds and leaves	small pox	Grinding	Paste is applied orally
Hadabanga	*Quadrang ula*		Barks	Bone fracture	Grinding	Barks and leaves paste is applied orally
Amla	*Emblica officinalis*	Euphorbiaceace	Fruits	use- vit-c, cough diabetes hyperacidity, cold	powder	Salts mixed of powder take
Tulsi	*Ocimum sanctum*	Lamiaceace	leaves, seed	cough, cold, bronchitis	Grind	50 ml applied for leaves taken orally
Grit kumara	*Aloe verra*	Liliaceae	Leaves	Skill burns, care, ulcer, wound healing	Grind	Latex paste and applied orally
Arakha	*Calotropis gigantean*	Asclepaidaceae	Leaves, fruits	snake bite, scorpion bite	Grind	Roots, leaves, latex are grinding paste and tied to the spot place with applied for daily 2 times take the medicine
Bisalya karani	*Tridax procumbens L.*	Asteraceae	Leaves	wound	Grind	Leaves paste is applied orally
Kusum	*Schleichera Oleosa*	Sapindaceae	stem, bark	fever, scabies, gout	Grind	Oil massage orally

Local Name	*Scientific Name*	*Family Name*	*Which Parts Used*	*Which Diseases*	*Purification Procedure*	*Preparation Procedure*
Harida	*Terminalia Chebula*	Combretaceae	Seed	wound, ulcer, leprosy	Powder	Seeds dry powder with paste applied orally
Bahada	*Terminalia Bellerica*	Comretaceae	bark, seed	cough, vomiting, ulcer	Powder	Seeds dry powder with paste applied orally
Dimiri	*Ficus racemose l.*	Moraceae	Fruits	lactation	Water boiled	Fruits is cooked and gives two diet
Mahula	*Madhuca indica*	Sapotaceae	Dry flower	fever	Water mixed	Fruit juices bring out of the purification wine drink one bottle
Patalagaruda	*Rauvolfia serpentina*	Apocynaceae	Roots	Snake bite	Grinding	Root paste is applied locally and evil spirits
Chandan	*Santaluna lbum*	Santalacea	seed/stem	Boil	Grinding	Grinding powder paste applied
Satabari	*Asparagus racemosus*	Liliaceae	Whole plants	Mensuration problems	Grinding	Grinding paste is given to drinks
Kendu	*Diospyros melanoxylon*	Ebenaceae	Fruits/seed	Dysentry	Grinding	Seeds and fruits paste taken
Palas	*Butea monsperma*	Fabaceae	Barks	Liver disorder	Grinding	A glass of latex and barks grinding orally
Gangasiuli	*Ocimuntenui florum*	Oleaceae	Leaves	Malaria	Grinding	Grinding the leaves and taken 50 ml in 8 days

plants continue to be there, be it for medicine or food or for material and religious culture people would continue to care and conserve such species from extinction. The survival of a culture requires integrity and creativity. The survival of ethno-medicinal culture requires that the dynamics that kept it alive over centuries be protected and cherished. This is the dynamic of the commons – the dynamic where the community sustains resources through use and conservation, the dynamic where the knowledge is free to grow and freely available to all.

Ethno-medicine is the mother of all other systems of medicine. Recently the importance of these traditional medicines has been realized worldwide as some of them proved to be very effective. Further scientific analysis of the plants in relation to the diseases on which they are applied would be required for pharmacological validation of these traditional medicines. The knowledge on the subject is gradually fading away because of lack of interest of local youth to go by the traditional knowledge of local folk healers. Hence, more organized efforts should be made to document traditional knowledge of the local people so in order to conserve the knowledge systems before they are gone with time.

So, it is suggested that in tribal area the knowledge of herbal medicinal plant should be introduced and tribals need to be educated about the medicinal plant value. The Government and Non-Government (NGO) agencies should take forward a step for preservation of the medicinal plants not only for the future health purpose of the world but also for the Health and Sanitation of the Hill Kharia Tribes.

REFERENCES

Anderson CR (2001) Your guide to health, Oriental watchman Publishing house, pp.644-712.

Asmit and Batra (2014) Ethnomedicinal plants of Mount Abu region in Rajasthan. World journal of science and pharmaceutical science 1(1): 28-37.

Bakshi SR and Bala K (1999) Development of Women, Children and Weaker Sections Social and Economic Development of Schedule Tribes. Deep and Deep Publications Pvt.Ltd, New Delhi

Basu SK (1993) Tribal Health in India. Manak Publishers Pvt. Ltd, New Delhi.

Chopra, Nayar, Chopra, (1956) Glossary of Indian Medicinal plants CSIR, New Delhi. Das, Bhagwan, Fundamentals of Ayurvedic medicine, Konark publishers, (Delhi, 1989)

Dastur JF(1988)Medicinal Plants of India and Pakistan. D. B. Taraporewala Sons Pvt. Ltd., Mumbai.

Jain SK (1981) Glimpses of Ethnobotany, National Book trust of India, A.S. Green Park, New Delhi.

Kalla AK and Joshi PC (2004) Tribal Health and Medicines. Ashok Kumar Mittal Publishing Company. New Delhi-110059.

Kar A and Choudhury BK (1994) Important mineral content of a few Ayurvedic herbs with a discussion on Medicinal aspects. Indian drugs 31(3): 127-130.

Khanna G (1975) All about herbal remedies, Vikal Publishing house, Pvt. Limited (New Delhi, 1994).Kloss, Jethro: Back to Eden, life line books, California.

Mohapatra B, Parida R and Jena MK (2016) An Assessment Of Folk Medicinal Use Of Plants By Tribes In Similipal, ADIVASI, (56) Vol.1, BBSR.

Nadkarni KM (1954) Indian Materia Medica, 3rd edition Popular Book Dep., Bombay.

Nagarjuna S.et.al (1982) Cultivation and utilization of medicinal plants, CSIR, Jammu Tawi, p.584-604.

Nayar SL and Chopra IC (1986) Research in Indigenous Drugs, In Glossary of Indian medicinal Plants, CSIR, New Delhi, I-XI

Reddy P, Adinarayana D, Uma Devi and Preethi V (2014) Inclusive Growth of Tribal Women Status and Contours. Sarup Book Publishers Pvt. Ltd New Delhi-110002

Roberto C (1988) The Mcdonald encyclopedia of Medicinal plants, Mcdonald and Company limited, London.

Roy Burman JJ (2003) Tribal medicine Traditional Practices and Changes in SIKKIM. Mittal Publications. New Delhi-110059

Sarkar RM (2008) Primitive Tribal Groups in India Tradition, Development and Transformation. Serials Publications. New Delhi

Schultes RE(1962) The role of Ethnobotanist in the search forthe new medicinal plants, Lioydia 25(4) : 257-266

Sharma AN (2007) Health Studies in Anthropology. Sarup and Sons New Delhi-110002

Chapter 7

Economy, Education and Survivability of Food-Gatherer-Hunters of Similipal Biosphere Reserve

Laxman Kumar Sahoo

Department of Anthropology and Tribal Studies, MSCB University, Baripada, Odisha
e-mail: laxman.sahoo2@gmail.com

Introduction

The present work is devoted to the study of economy, education and survivability among the so called Particularly Vulnerable Tribal Groups (PVTGs) like the Hill Kharia and the Birhor(*Mankirdia*). All of whom belong to Mayurbhanj district and few other adjacent areas of North Odisha. Their living place is Similipal Biosphere Reserve. Earlier these two tribes were categories as primitive tribal groups (PTGs) which have been replaced by the term Particularly Vulnerable Tribal Groups (PVTGs) in 11th five-year plan period. The two tribes under study are economically backward and particularly vulnerable, their major time in a day is spent for food quest and they avail very little time for other extra cultural activities. Therefore, the state of their economy and problem of formal education as well as survivability has been studied.

Hill Kharia

The Hill Kharias of Odisha are mostly confining to the Mayurbhanj district of Odisha. Sometimes they are confused with the Kharia of Sundargarh who are known as "Dudh" and "Dhelki" Kharias. In Mayurbhanj district Hill Kharias are simply known as Kharia and prefix Hill is used only in official expression. However, during the pre independence time they are known as "Erenga" Khadia living mostly at the out skirt of the Similipal forest range. As most of the Kharias say they were living with a fort or Gada in Jashipur which was named after Jashu Kharia, the chief of the community. It is said to be more than a century old mound where the Hill Kharia chief Jashu had his residence. He was a land lord and Kharias were cultivators at the out skirt of Similipal forest range. As per the local saying, they were defeated by the Bathudi community who pushed them inside the forest. Inside the Similipal forest Khairas were compelled to live on the plant and animal foods and because of long term dependence they were denoted as hunter-gathers with a backward economy. As regards the term Kharia, in Jharkhand and its adjacent areas Kharias are said to be the palanquin-bearers of the kings and Jamindars and in the locality such palanquins known as "Khar Kharia". Because of which they are called as Kharias later. Because of the shift from the forest based economy to agricultural economy and further coming back to the same forest based economy they have been said to undergo the process of secondary Primitivization which has been vividly discussed by the well-known anthropologist Surajit Chandra Sinha (Dash,1998). Because of the economic backwardness mostly depended on the plant and other resources of the forest, they are considered to be one of the 13 PVTGs of Odisha. During the seventh five year plan (1985-90), they have been included under a micro project along with Mankirdia/Birhor. The micro project is thus named as Hill Kharia Development Agency located at Jashipur of Mayurbhanj district. It was established in 1986 by including the Khadias of Khejuri, Barhakamuda, Gudugudia are included under the micro projects and several others in the eastern part of Similipal are surprisingly excluded from the list. Birhor/Mankirdia of only Kendumundi and Dhobadiha settlements are included. As other Birhors leads a nomadic life around Similipal forest range are excluded from the micro project.

Economy

The economy of a tribe is the projection of its socio-economic system. It demonstrates the level of technology a tribe possesses, and the manner in which the tribes has responded to the eco-system in which it placed; the distribution of goods and services is regulated by considerations of right, obligation, reciprocity and certain mutuality. According to Paul Bohanan (1963), "economy is the way in which resources technology and work are combined to satisfy the materials requirement of human beings and social groups."

In the district of Mayurbhanj, the Kharias of Similipal hills are only known as Hill Kharias. In the Sadar subdivision of Mayurbhanj district there are several Khadia families who lives far away from Similipal and depend on agricultural economy.

The so called Hill Kharias in and around Similipal Hills though started depending on agricultural economy by encroaching the forest land, their primary occupation was hunting and gathering of plant and animal resources of the forest ecosystem. Usually Hill Kharias maintain several seasonal collections of food and non food items from the forest out of them three items are economically very important for the maintenance of livelihood. They are honey, arrow root and Sal resin from the forest. Whereas honey is collected throughout the year, arrow root is collected only during winter season and Sal resin is collected towards end part of rainy season and beginning of winter season. Sometimes also Sal resin collected during summer season. Usually these three major collections are collected mostly for the market. In the past marketing was mostly organised by the forest department when people as selling to the forest department at a lower price as a result of which the major part of the resources are mostly sold to the local traders mainly because they give advance at the time of need and the Khadias are obliged to sell them their forest collections. Besides these three Major collections Hill Kharias also collect varieties of mushrooms, edible leaves, edible and medicinal flowers, broom stick, natural ochres, tassar cocoons, Varieties of gums and several animal products. In all these collections they maintain a very strict division of labour. Generally women collect plant foods like roots, tubers, fruits, flowers, leaves, *etc.* from the forest. Whereas the male members they organise hunting of game animals, honey collection, Sal resin collection and partly arrow root collection. Usually Hill Kharias start their forest expeditions only after Magh Puja in the month of January and February. They propitiate all the forest deities to avail their blessings for the collection of required resources around the year.

Education

While addressing the problem of tribal education, it cannot be said that all tribes are facing similar problems equally. Some tribes like Santal, Kol, Bhumij, Bathudi *etc.*. of North Odisha are relatively more acculturated and dynamic. Though not all, majority of their children are coming to the school with personal interest. Usually it is found that vulnerable tribes are not coming up in the field of formal education. Mainly because they maintain subsistence or below subsistence level of economy, all the family members including the school going children have to join hands in the productive or economic activities of the family. Specifically in case of Hill Kharia, a semi-nomadic community around Similipal hills, Mayurbhanj District, the situation is relatively more precarious because of their primary dependence on food -gathering activities in the forest. Because of semi nomadic, some part of the year especially during the time of arrow root collection and food gathering activities, all the family members stay inside the forest for the gathering of plant foods and other requirements. This results in the deprivation of primary school education of the children.

After the declaration of Similipal Hills, a Tiger Reserve in 1973, Hill Kharias inhabiting the core areas are displaced out of the forest area. Due to the absence of natural or forest resources in the sites of rehabilitation, they are compelled to be daily wage laborers for their basic minimum subsistence. Even though there are school facilities in the rehabilitated host villages (namely Banabasa) near Jashipur, poverty has debarred them to go for school education. However, mainly because of mid day meal programme, some such Hill Kharia children feel attracted towards the school but because of their full time association with the subsistence activities of the family, they are not able to maintain the continuity. It results in the increase of the cases of absenteeism and drop outs.

Birhor (*Mankirdia*)

The Birhor are only nomadic tribe of the state they are said to be the inhabitants of Jharkhand who in course of their migration came to Odisha and confine to similal forest ecosystem. They are known in different names in the state of Odisha according to their popular practice of monkey hunting and catching. Their original name is Birhor which is a combination of "Bir+hol". Whereas 'Bir' refers to forest and 'hol' refers to man. In local Odia people call the monkeys as mankad because of which the Birhors are called by the tern mankidia or mankirdia because of their monkey hunting (Dash,1998).

Birhors live in four adjoined states namely Bihar/Jharkhand, West Bengal, Odisha and Chhattisgarh. They speak a dialect under Mundari group to same extent similar to that of Santal and Kol. Because of their nomadic habit they mostly live in small bands comprised of ten to fifteen nuclear families which are known as "Tanda" in Mundari language. The families are mainly of patrilineal exogamous group. In the "Tanda" or bands there exists a girl's dormitory and where the unmarried girls of the 'Tanda' live together.

Economy

The Birhor economy is mainly depended on forest resources. They are primary hunter-gathers with a nomadic way of life. They always stay in between the resourceful forest and the peasants by constructing temporary huts at the outskirt of a peasant village may be tribal or non tribal. They mainly depend on the rope making out of the wild creepers. Out of several creepers which are used for rope making the best one is Siali (*Bauhinia Vahlii*). They usually collect barks of the matured creepers for about five days in a week and for the next two days they prepare rope out of it and sell them either in the weekly market or in the peasant villages both for cash and kind. This is major source of income for the Birhors. Besides they also collect lot of edible and inedible resources from the nearby forest. As they mainly depend on Siali creepers of the forest for earning cash, the moment creepers are exhausted from the local forest they shift to another forest area for the resource collection. In a particular place they usually do not stay more than two three months. Of course the duration of stay is mostly dependent on the availability of Siali creepers in the forest. Though the Birhors are not good hunters like the Santals and other neighbouring community they are quiet specialized in monkey hunting and catching which no other tribe follow. They always maintain the great desire for relishing the flesh of Monkeys unlike other ethnic groups in Odisha. Whenever they collect good amount of Siali barks and other useful resources from the forest they specially plan for monkey hunting once or twice in a fortnight. The hunting process is very systematic. On the day of hunting, all the male members of the 'Tanda' proceed to the forest and search for the monkey troop in a specific site then divide into four groups, on group will chase the monkey towards the neck set on the ground little away in an open field, the next two groups will guard both the sides not to allow the monkey to escape and the fourth group hide themselves at the net. When the chasers from the back side and both sides chase the Monkey troop, the Monkeys try to go in a forward direction and near the open field they climb down from the tree and run. At the end part of open field, when they run being terrified, they are cut at the net and immediately the hidden group members stand up and strike them on the head. The monkeys after cut or hunt, are carried to the Tanda and distributed among the families in a very systematic manner. In the forest of Similipal hills mostly they hunt two types of monkeys which they call as hanu (*black faced*) and gali (*red faced*). After the flesh is consumed the hides are sold to the local people to make musical instruments out of them.

As a hunting and gathering community, Birhors follow a definite division of labour both in gathering and hunting activities. Early in the morning all the adult males and females after taking watered rice start for the forest throughout the day the male members collect Siali fibres and females mostly other edible and inedible resources required for the family. In the afternoon they all comeback and all the forest collections are usually individually owned but they move in a group all the time for the sake of security in the forest. However, the hunting activity (*Sandra*) is always performing collectively.

Education

The problem of education among the Birhor, a hunting and gathering nomadic tribe has been a great concern for the policy makers and development workers. The Birhors are the only nomadic tribe of the state who are still continuing an archaic mode of living completely based on forest resources. Because of nomadism, such people do not stay in a particular place for not more than three months at a stretch. They live in a temporary hut without any minimum facility of drinking water, electricity and sanitary requirements. Looking at the availability of natural resources in the nearby forest, they decide to stay one or more months in a particular place. However, they usually prefer to stay near a peasant village for getting food grains, water *etc.*. and marketing of their forest products. Though, they collect variety of forest products, they give prime importance to the collection of 'Siali' fibers (*Bauhinia vahlii*) mainly for rope making. In every week or fortnight they also prefer to go for monkey hunting in the forest.

With reference to the above stated nomadic ways of living of the Birhor, it is very difficult to think about formal education. Because of their constant mobility and primary dependence on the collection of forest resources, it is extremely difficult to bring their children to primary education. However, the State Government of Odisha has been successful in bringing some of them to settled-living in two different Indira-Awas colonies, located at Kendumundi in between Karanjia and Thakurmunda, and at Durdura near Jashipur. Because of settled life some of the Birhor children in the said colonies are going to the nearby schools being attracted by mid-day meals.

But the educational programme has not been successful very much mainly because Birhors here still follow forest dependent activities and often maintain semi nomadic movements in Similipal hill range. This has resulted too much absenteeism and dropouts among the Birhor children of the school.

Question of Survivability

The concept of 'subsistence' primarily deals with the means of existence. The meaning of 'Subsistence' is the means for supporting of life; a living or livelihood of any living species including human beings. When the theory of subsistence mingles with the concept of claim, a substantive image comes into light which is known as the 'Subsistence Rights', which emphasises the most essential claim for existence or survival. The Subsistence Rights are these inalienable rights available to the mankind since the civilization dawns irrespective of caste, creed, religion, sex, status and social position. Subsistence Rights are the minimal right which every individual must have against the State or other public authority by virtue of his being a 'member of human family', irrespective of any other consideration. These rights are inevitable for the development of human society as a whole, as well as the individual development of human being.

As Similipal has been declared a National Park as well as wild life sanctuary since 1958 and a tiger reserve since 1973, restrictions have been imposed on the collection of minor forest produce in the reserve forest area and hunting has been strictly prohibited both in the reserve and protected forests. The non-tribal exploits them beyond their basic need, mostly for their commercial benefits. The forest-based community, forest management has become a vehicle for self-determination for ethnic minorities. Among the tribal group, management is highly participatory but most excellence are reflected in the traditional system have developed over centuries-and embraced. The role of culture for their development and continued maintainers is pervasive. In fact, the revival and preservation of traditional management practices does not only go hand-in-hand with maintaining cultural integrity and identity; but they form a symbiosis, the survival of which is as crucial for the people as it is for the forest they manage. A struggle to preserve traditional property rights is also hallmark in these communities. An important point to think twice is before tinkering with something's that has worked for generation. It is most likely outstanding-otherwise it would have long been extinct. It has been realised that the protection of biodiversity save the future of human long run. It is not only the number and spread the species, but also the presence of genes or characters for resistance to diseases and pathogens, and a variety of environmental stress reside in the wild plants.

In recent years the interaction between tribes and their environment has been subject of great concern for the policy makers and development practioners. The environment in which they adapt themselves and which they also mould includes not only the physical aspect but also socio-cultural and biological aspects. The environment comprises all over the surroundings of any object. Movement of tribes from one place to another, merciless exploitation of the resources without any element of conservation or development and growth or environment/forest laws have caused ecological imbalance and which in turn threaten the very survival or reproduction of living organism. The Birhor are excellent monkey catchers. They eat monkey flesh and trade with monkey skin. But in the Wildlife Protection Act 1972, not to do such type of animal hunting activities, as a result the Birhor found themselves nutritious deficiencies and food crisis. Historically the terms Work and Labour have been used synonymously and interchangeably as both 'Work' and 'Labour' requires human effort *i.e.*, an exercise of body and mind. Although work is directly fulfilling the child's natural abilities and creative potentialities, always conducive to his healthy growth, but when it is meant for the fulfillment of other's needs, becomes enslaving in character and deleterious in impact. When the basic attributes of work *i.e.*, purpose, plan and freedom are conspicuously absent, it becomes labour. In this way the Birhors and Hill Khadias of Similipal hills bound to accept migrant labourer is an alternative economy.

There are quite a few tribal groups that constitute slightly more than one per cent of the total population namely HO (1.42) in Eastern India and Naga (1.23) in the North Eastern India states. Apart from them, there are numerous smaller tribal groups dispersed across the country. In fact, there are some 'small' so-called vulnerable tribes of which enumerated populations range from as low as 20. Ignoring possible biases both in enumeration and complication of enumerated data from censuses, it is clear that many such tribes are indeed on their way towards extinction. This often leads to an alarmist note on so-called vanishing tribes articulated mostly by anthropologist who's professional and academic affinity chiefly lies in micro level studies of small tribal communities. It should be noted, however, that such a phenomenon of shrinking population is often localized and specific to a very small group of tribal people who are situated in very special circumstances (Maharatna,2005). These most vulnerable tribal groups, who are often branded as PVTGs, currently number around 75 and constitute nearly 2 per cent of the total population which includes, Onges, Shompen, the Greater Andamanese, Hill Khadias, Birhor and Bonda. To arrest the phenomenon of declining population size among these small tribal groups and to rehabilitate them is a challenging task, but little more than a mere recognition of this problem has so far been forthcoming. Example like the shrinking tribes (Hill Khadia and Birhor) of the Similipal hills have often been attributed to specific genetic factors that have been nurtured by complex and long standing interactions between geophysical features, tribal seclusion and forces of modernization. Acute deprivation and extreme vulnerability to death and diseases are responsible for the declining numbers of small tribal groups of people in specific location.

Conclusion

Tribal subsistence and survivability in entire India as well as in Mayurbhanj district of Odisha is basically forest based. In Mayurbhanj district the tribal subsistence has been supported and influenced by the massive forest range of Similipal hills. Tribal communities and natural wealth are inseparable entities; one cannot survive and thrive without the other. The most successful economic pursuit man had adopted was food gathering and hunting, activities for the most minimum living. Similipal has a greater magnitude of bio-diversity richness to fulfill all sorts of demands for subsistence/survivability of the tribal people. There are several issues like shrinking of wildlife habitation area, annual mass hunting traditional ritual (*Akhanda Shikar*) of the tribals, cattle grazing, mining, growth of human population and rise of wildlife crime are responsible for the loss of naturalness of Similipal. In spite of all endeavours through the forest acts and laws, the forest resources are undergoing fast depletion. Consequently they adopt marginal agricultural practice and wage and migrant labour to manage their livelihood. Above all many legal provisions enshrined in various forest legislations are stand as obstacles in the path of the subsistence of tribals.

Hill Kharia and the Birhor are simple communities and they are not uneducated since they acquire the required knowledge in act depending on their cultural type. In such simple societies people carry on various socio-cultural activities as per their cultural standards in the process they acquire experience and knowledge through the process of enculturation. The experience and knowledge is mostly denoted as a survival strategy for the concerned society. On one hand such knowledge and experience are mostly exercised for the purpose of livelihood maintenance but on the other hand it may be depicted as a formal educational system materialized through the socialization and enculturation processes.

Under the impact of State sponsored economic development programmes and processes of modernization traditional cultures of tribal communities have started changing. Tribal socio-economic development and forest development cannot be separated without hearing the interests of tribals and the forest in the country. Only after achieving such knowledge, we can look forward to a better future.

REFERENCES

Bohannan P (1963) Social anthropology. Holt, Reinhart and Winston, New York.

Dash J (1998) Human Ecology of Foragers: A Study of the Kharia (Savaras), Ujia(Savara) and Birhor in Similipal Hill, Commonwealth: New Delhi.

Maharatna A (2005) Demographic Perspectives on India's Tribes. Oxford University Press, New Delhi.

Chapter 8

Navigating Menopausal Transitions: A Focus on Tribal Women in Mayurbhanj, Odisha

Devaranjan Das[1]*, Subhashree Panda*[2] *and Gargee Mohanty*[1]

[1]*P.G. Department of Zoology, Maharaja Sriram Chandra Bhanja Deo University, Baripada, Mayurbhanj, Odisha*
[2]*Parala Maharaja Engg. College, Sitalapalli, Berhampur, Odisha*
e-mail: gargee.mohanty@gmail.com

Introduction

Women, the epitome of natural beauty and the wellspring of life's inspiration, navigate a journey marked by transformative phases and challenges. Menarche, heralding the onset of the reproductive cycle, unfolds between 11 to 16 years, while motherhood emerges as a breathtaking chapter in a woman's life. Menopause, though, stands as the last, often disheartening phase, causing both physical and mental upheaval. From irregular menstrual cycles to joint pain, fatigue, and psychological shifts, menopausal symptoms are diverse. The term "menopause" itself, derived from the Latin words 'Meno' (month) and 'Pause' (stop), signifies the complete cessation of menstruation for over a year, primarily due to ovarian function decline (Gold *et al.,* 2011).

The stages of menopause, as categorized by the Stages of Reproductive Aging Workshop (STRAW), encompass pre-menopause, peri-menopause, menopause, and post-menopause (Rahman *et al.*, 2010; Harlow *et al.*, 2012). Pre-menopause initiates with menarche, extending through a woman's 40s and 50s, culminating with the first signs of physiological or psychological transformation. Peri-menopause, the transitional period around menopause, witnesses hormonal fluctuations, impacting physical and mental well-being. Post-menopause, identified by a year-long absence of menstruation, marks the loss of reproductive capability due to hormonal changes (Khatoon *et al.*, 2018).

Directing attention towards the tribal communities of Odisha which highlights the paramount importance of health awareness, particularly for women grappling with menopausal issues. Mayurbhanj, is one of the largest and tribal districts of Odisha. This district predominantly inhabited by tribal communities and characterized by rich cultural diversity, wrestles with the consequences of modernization, contributing to a surge in lifestyle-related diseases within its population (Mohanty, 2017). Given this scenario, addressing health concerns becomes not only crucial but also an urgent necessity. In this specific context, gaining insights into the health status of peri and post-menopausal tribal women emerges as a pivotal stride. Unfortunately, the absence of systematic data and awareness further underscores the pressing need for tailored healthcare interventions in this often-overlooked segment of society.

The present review of the subject mentioned thus aims to shed light on the challenges faced by peri- and post-menopausal tribal women in Mayurbhanj district, Odisha. Specifically, the investigation probes into the physical and mental health, nutritional, and lifestyle issues affecting these women, aiming to uncover their prevalence. The outcomes of this review work are poised to offer valuable insights, enabling the government and other societal stakeholders to formulate improved policies and establish enhanced healthcare systems catering specifically to the needs of elderly women in tribal communities.

Definition of Menopause

The term "menopause" was originally coined by Gardanne in 1816 (Wilbush, 1979). Menopause is characterized by the complete cessation of the menstrual cycle, marking the end of a woman's reproductive age. Initially, menopause was attributed to a deficiency in ganglionic regulatory functions. However, subsequent research revealed that the ovaries play a pivotal role in the onset of menopause. Specifically, ovarian cells cease to produce optimal levels of estrogen and progesterone, disrupting the menstrual cycle and triggering menopause (Marshall, 1910; McKinlay *et al.*, 1985).

The age at which menopause occurs varies among individuals, with menopausal symptoms typically manifesting in the late 30s. Predicting the exact date of menopause is challenging, as the official designation is established twelve months after the final menstrual cycle (WHO, 1996a).

Average Age of Menopause

Average Age at Menopause on the International Level

Globally, the median age at menopause typically ranges between 45 and 55 years (Kaufert and Syrotuik, 1981). Notably, women in Western countries and some Middle Eastern nations tend to experience menopause at a later age compared to counterparts in other parts of the world (Khatoon *et al.,* 2018). The mean age of menopause for Italian women is reported to be 50.9 years (Meschia *et al.,* 2000), while in Iran, it is documented at 50.4 years (Mohammad *et al.,* 2004). In Slovenia, the average age of menopause is found to be 50.4 years (Sievert *et al.,* 2004), and in Korea, it is reported as 49.3 years (Ku *et al.,* 2004). On the lower end, Turkey has a relatively low average menopausal age of 45.8 years (Biri *et al.,* 2005), whereas Singapore records an average of about 49.1 years (Chim *et al.,* 2002). Taiwan reports an average age of 48 years (Fuh *et al.,* 2001), and Malaysia records an average age of 51 years (Rahman *et al.,* 2010; Khatoon *et al.,* 2018). In summary, the data reveals a lack of uniformity in menopausal ages across different countries, reflecting a diverse global distribution.

Average Age at Menopause at the National Level in India

The mean age at menopause for Indian women is reported to be 44.3 years, as per the National Family and Health Survey - 2 conducted in 1998-99 (NFHS-2, 1998-99). Approximately 11.0 per cent of women experience menopause before the age of 40 (Syamala and Sivakami, 2005), and this frequency has slightly increased by 0.5 per cent in subsequent years (IIPS and ORC, 2007). Studies conducted in India indicate that the age at menopause among Indian women varies due to factors such as diverse ethnic and socio-economic groups, women's working status, and the rural-urban divide (Dasgupta and Ray, 2009; Nagar and Dave, 2017; Sidhu *et al.,* 2017).

The mean age at menopause for specific ethnic groups in India is as follows: Hindu Kalita women (44.94 years) (Bagga, 2004), rural North Indian women (44.10 years) (Singh and Arora, 2005), Punjabi Brahmin of Punjab (48.22 years) (Kaur and Talwar, 2009), Kshatriya of Andhra Pradesh (46.04 years) (Kapur *et al.,* 2009), North Indian women from Jammu (47.5 years) (Sharma *et al.,* 2007), Bengali Hindus in West Bengal (46.9 years) (Dasgupta and Ray, 2009), and Gujarati women in Vadodara (47.7 years) (Christian *et al.,* 2011).

Several studies have attempted to identify determinants of age at menopause, but the results are not consistent (Kriplani and Banerjee, 2005; Dasgupta and Ray, 2009). For instance, a study on North Indian women found no significant effect of sociodemographic and reproductive factors on the age at menopause (Kriplani and Banerjee, 2005). However, a study from Eastern India revealed that the duration of breastfeeding and rural-urban habitats is significantly associated with the age at menopause (Dasgupta and Ray, 2009).

Stages of Menopause

Determining the stage of menopause is crucial for numerous epidemiological studies that aim to establish relationships between menopausal symptoms, anomalies, and their incidence across different phases of life and populations. Authorities have commonly classified menopause into pre-menopause, peri-menopause, and post-menopause. Various interpretations exist; Jaszmann *et al.* (1969), Thompson *et al.* (1973), and WHO (1996b) classify post-menopause based on the Last Menstrual Period (LMP), considering twelve months since the last menstrual period. Belchetz *et al.* (1973) suggest six months from LMP, while Winokur (1973) proposes three years after the last menstrual period. Neugarten and Kraines (1965) advocate for self-reporting as an authentic means of menopausal classification.

Dudley *et al.* (1998) and Mishra *et al.* (2003) present separate definitions for peri- and post-menopause. Peri-menopausal status is attributed to women experiencing irregular menstrual periods within the twelve months preceding the last menstrual period or a complete absence of menstrual bleeding for more than three months but less than twelve months. Post-menopause is defined for women who have naturally ceased menstrual bleeding for at least twelve months before the last menstrual period.

A more recent classification by Harlow *et al.* (2012) provides distinct definitions for different stages of menopause. In their workshop themed on the stages of Reproductive Aging Workshop, they categorize early menopause transition as stage-2, characterized by inconsistency in the length of the menstrual cycle. This stage involves cycles recurring with a time difference of seven days or more between two successive cycles. Late menopause, denoted as stage-2, is established by amenorrhea lasting sixty days or more. Early menopause stage spans approximately five to seven years from the occurrence of the final menstrual period. Similarly, late post-menopause, termed stage +2, occurs five years after the final menstrual period, during which further changes in reproductive hormonal functions are limited, and more of the physical and physiological aging process occurs (Harlow *et al.*, 2012; Khatoon *et al.*, 2018).

Hormonal Changes during Menopause

Menopause signifies notable hormonal fluctuations that affect a woman's reproductive system. With advancing age, the ovaries gradually reduce their production of estrogen and progesterone—key hormones crucial for the menstrual cycle and reproductive processes.

The perimenopausal phase, preceding menopause, often exhibits irregular menstrual cycles due to fluctuating hormone levels. As menopause nears, the ovaries cease releasing eggs, causing a decline in estrogen and progesterone production. Estrogen, specifically, plays a vital role in regulating the menstrual cycle and maintaining reproductive tissue health.

Reduced estrogen levels can lead to symptoms such as hot flashes, night sweats, mood swings, and vaginal dryness, indicative of the body adjusting to the changing hormonal landscape. Additionally, the decline in estrogen during menopause may contribute to long-term health effects like osteoporosis and an increased risk of cardiovascular diseases. Hormone replacement therapy (HRT) is sometimes used to alleviate menopausal symptoms by supplementing estrogen levels, though its risks and benefits require careful consideration.

Understanding the hormonal changes during menopause is crucial for managing associated symptoms and addressing potential health risks. Regular medical check-ups and consultations with healthcare professionals can assist women in navigating this transitional phase with greater comfort and well-being.

At birth, both ovaries contain approximately one to two million primordial follicles, each consisting of a single layer of granulosa cells. By puberty, about 0.4 million follicles remain, with naturally developing follicles stimulated monthly by follicle-stimulating hormones. About two weeks after the menstrual cycle, one follicle matures, producing the majority of estradiol and ovulating due to Luteinizing Hormone. The follicle transforms into a corpus luteum, releasing progesterone along with estradiol.

Around a week after ovulation, the corpus luteum peaks, then retreats, leading to decreased hormone production and subsequent menstrual bleeding around fifteen days after ovulation. As women age, the number of ovarian follicles drops, and follicular activity decreases after 40 to 50 years. This results in increased synthesis and release of FSH, lowering estradiol and inhibin levels. Four or five years before menopause, menstrual cycles become irregular due to shorter or longer ovulatory cycles. Some follicles lose quality, leading to irregular anovulatory cycles. Menopause marks insufficient estradiol production to trigger endometrium, resulting in the cessation of menstrual bleeding and the decline of fertility in peri- and post-menopausal women.

Symptoms of Menopause

Menopausal-related symptoms can be classified into three types: vasomotor symptoms (VMS), known as somatic symptoms, psychological symptoms, and urogenital or sexual symptoms (Rahman *et al.,* 2010; Khatoon *et al.,* 2018). The assessment of the prevalence of these symptoms in women at different stages of menopause is commonly conducted using the Menopause Rating Scale (MRS), a widely utilized tool for such investigations.

Within the domain of somatic symptoms, physical manifestations of menopause are encompassed. Hot flashes and sweating, characterized by sudden sensations of heat, are common experiences. Heart discomfort, including palpitations or irregular heartbeat, may occur. Sleeping problems, such as difficulties falling or staying asleep, contribute to somatic symptoms. Additionally, joint and muscle discomfort may be present, along with sexual problems impacting sexual function.

The psychological domain assesses emotional well-being during menopause. Depressive mood, marked by feelings of sadness, and irritability, characterized by being easily provoked, are common psychological symptoms. Anxiety, a sense of unease or worry, may also be experienced. Physical and mental exhaustion, often described as overwhelming fatigue affecting both body and mind, is another psychological symptom. Sleeping problems, contributing to disruptions in sleep patterns, fall within this domain as well.

Urogenital symptoms pertain to issues in the urinary and genital areas. Sexual problems, affecting desire or performance, are assessed within this domain. Bladder problems, encompassing issues related to urinary function, are considered. Vaginal dryness, characterized by dryness, itching, or discomfort in the vaginal area, is another urogenital symptom.

Vasomotor or Somatic Symptoms

Hot flashes and night sweats are thermoregulatory instabilities that commonly characterize menopause. During hot flashes, women experience a sudden sensation of heat in the face, neck, and chest regions, often accompanied by a higher rate of heart palpitations (WHO, 1996a). These episodes can be triggered by tension and anxiety (Tepper *et al.,* 1987) and cause distress in approximately 60 per cent to 80 per cent of women entering menopause (Freeman and Sherif, 2007).

Hot flashes result from ovarian irregularities during the peri- and post-menopausal periods. In post-menopausal women, these flashes are linked to a disturbance in the temperature-regulating mechanism in the hypothalamus, attributed to the declining levels of estrogen due to the diminishing functionality of ovarian follicles.

As the name suggests, "Night Sweats" occur during nighttime and resemble daytime hot flashes. These episodes start with a sudden feeling of heat on the face, neck, upper back, and chest areas. A majority of women experience hot flashes and night sweats during their menopausal transition (Gold *et al.,* 2006; Thurston and Joffe, 2011).

Normally, the body maintains its temperature through vasodilation and sweating in response to heat and shivering when feeling cold. However, in menopausal women, the "thermoneutral zone" narrows down, causing even a slight change in temperature to trigger a regulatory response through sweating or shivering (Freedman, 2005).

Various factors contribute to the occurrence of hot flashes, including minimal physical activities, obesity, smoking, alcohol consumption, warm ambient temperature, and hot drinks. The incidence of hot flashes varies among ethnic groups, with reports indicating that Caucasian women experience more hot flashes compared to Japanese and Chinese women. Similarly, Japanese and Chinese women exhibit more frequent hot flashes than African-American women (Gold *et al.,* 2006).

Psychological and Other Symptoms

The relationship between low estrogen levels and the onset of psychological symptoms in menopausal women remains a subject of controversy. A study conducted by Nicol-Smith (1996) did not establish a clear cause-and-effect relationship between lower estrogen levels and psychological symptoms. However, another investigation by Comings *et al.* (1999) found an induced effect of the estrogen receptor gene on the onset of anxiety in menopausal women. Subsequent studies also indicated that estrogen modulates serotoninergic function, contributing to mood swings in menopausal women (Joffe and Cohen, 1998).

While Joffe and Cohen (1998) linked low estrogen levels to the cause of mood swings, the topic remained debatable for years due to unresolved methodological issues in defining menopausal stages, measuring mood, and considering factors like age and social changes (Matthews *et al.,* 1990; Kaufert *et al.,* 1992).

Studies revealed that changes in estrogen levels are not the sole cause of depression and anxiety in menopausal women. Various factors, such as family status, upbringing, socio-cultural events, negative attitudes towards aging, and changes in family and societal roles, also contribute to secondary symptoms like anxiety and depression (Hunter, 1996; Dennerstein *et al.,* 1999; Bromberger *et al.,* 2001; Melby *et al.,* 2005; Woods *et al.,* 2006).

Urogenital Symptoms

Women experiencing vaginal and urinogenital organ atrophy often report discomfort, dysuria, dyspareunia, and urinary tract infections (UTIs), with these symptoms being more common after menopause (Milsom and Molander, 1998).

In human females, the urethra is estrogen-sensitive, with estrogen receptors found throughout the urinogenital tract. Low estrogen levels during menopause result in symptomatic cytological changes in the urinogenital tract, causing mucosal thinning and increased neutrophil infiltration (Reid *et al.,* 2014).

Additionally, the vulva undergoes changes, losing collagen molecules and adipose tissues due to decreased estrogen secretion. This leads to thinning and flattening of tissues, reduced water retention ability, loss of glandular secretion, and sensations of irritation and itching during activities like sitting and sexual activities (Oriba and Maibach1).

The loss of estrogen alters the flora of the urethra and vagina, making the tissue more basic, which, in turn, facilitates the colonization of enteric bacteria. This change in pH contributes to itching and irritation in the urinogenital organs (Iosif and Bekassy, 1984).

Cohort studies with Western populations reported that 27 per cent of the study group experienced vaginal dryness, 18.6 per cent reported itching and irritation, 11 per cent had vaginal discharge, and 5.2 per cent reported dysuria (Pastore *et al.,* 2004). Another population-based study indicated that symptoms related to vaginal dryness, itching, irritation, and others occur some years before and after the menopausal transition. A study by Dennerstein *et al.* (2000) found that 21 per cent of women experienced vaginal dryness within one year of menopause, increasing to 47 per cent by three years after the transition. Vaginal dryness was more prevalent in smokers than non-smokers (Kalogeraki *et al.*, 1996).

Problems associated with the urinogenital tract, particularly vaginal itching, dryness, and irritation, significantly impact the quality of life (Stenberg *et al.*, 1996). An anonymous survey conducted by Nappi and Kokot-Kierepa (2012) with 3520 women aged between 55-65 years worldwide found that almost 45 per cent of women reported experiencing various types of vaginal problems.

Measurement of Menopausal Symptoms

The assessment of the prevalence of menopausal symptoms in women at different stages is commonly conducted using the Menopause Rating Scale (MRS), a widely employed tool for such investigations. The MRS enables healthcare professionals to gain insights into various aspects of a woman's experience during menopause by categorizing symptoms based on their impact within each domain and assessing severity on a scale. This tool aids in comprehending the overall effect of menopausal symptoms on an individual's well-being and quality of life, offering valuable guidance for tailoring interventions to address specific needs.

The MRS is a standardized, self-administered questionnaire designed to assess the severity of menopausal symptoms across three domains: vasomotor, psychological, and urogenital. Within each domain, specific questions related to symptoms are presented, and scoring is carried out on a scale of 0 to 4, where 0 denotes the absence of the symptom, and 4 indicates 'very severe.' For simplicity, Yes/No answers are recorded as 1 and 0, respectively, streamlining the scoring process. Available in multiple languages, the MRS facilitates a comprehensive evaluation and monitoring of menopausal symptoms by healthcare professionals. Regular assessments using this scale provide valuable insights for tailored interventions, ultimately enhancing the quality of care for women navigating the complexities of the menopausal transition.

Menopausal Symptoms and their Determining Factors

There are many factors which are responsible in causing the onset and severity of menopausal symptoms. Here we shall be discussing some of such determining factors that aids in spiking the severity of menopausal symptoms.

Hormonal Changes

The intricate hormonal transformations occurring during menopause play a pivotal role in giving rise to various symptoms that impact women's physical and emotional well-being. The central actors in the reproductive processes, estrogen and progesterone, undergo notable fluctuations and declines during this transitional phase, contributing to a diverse array of physiological and psychological changes.

Menopause signifies the cessation of menstrual cycles, primarily triggered by the diminishing levels of estrogen and progesterone. Estrogen, a critical hormone produced predominantly in the ovaries, takes center stage in regulating the menstrual cycle and maintaining reproductive tissues. The aging and reduced functionality of the ovaries lead to a gradual decline in estrogen production, marking a hallmark of menopausal hormonal changes. Progesterone, another ovarian hormone crucial for preparing the uterine lining for potential pregnancy, experiences reductions as ovarian function diminishes, contributing to alterations in the menstrual cycle (Santoro *et al.,* 1999).

The decline in estrogen and progesterone levels results in various physiological consequences. A prevalent manifestation is the irregularity of menstrual cycles during the perimenopausal phase, characterized by erratic hormone levels and occasional anovulatory cycles (Reed *et al.,* 2008). Fluctuations in hormonal levels give rise to vasomotor symptoms such as hot flashes and night sweats. Hot flashes, sudden sensations of heat often accompanied by sweating, are linked to the body's thermoregulatory response to hormonal changes.

Estrogen's role in modulating neurotransmitters contributes to mood swings, irritability, and changes in emotional well-being. Some women may experience symptoms of anxiety or depression during the menopausal transition. Estrogen receptors, present throughout the urinogenital tract, undergo decline, leading to urogenital changes. Vaginal dryness, itching, and discomfort may occur due to a thinning of the vaginal mucosa, while changes in the urethra can contribute to urinary symptoms (Maki *et al.,* 2008).

Understanding these hormonal changes is crucial for addressing menopausal symptoms. While hormonal shifts are a natural aspect of aging, managing symptoms and promoting overall well-being during this transition often involves a combination of lifestyle adjustments, medical interventions, and emotional support. Regular communication with healthcare professionals facilitates navigation through this significant life stage with greater comfort and understanding.

Genetic Predisposition

Genetic predisposition plays a significant role in shaping the onset and severity of menopausal symptoms. Although menopause is a shared biological experience among women, the diverse ways individuals encounter and manage its symptoms exhibit considerable variation, with genetics playing a substantial role in this diversity.

The regulation of hormones, particularly estrogen and progesterone, is subject to genetic influence. Variations in genes associated with hormone receptors and metabolism can impact the delicate hormonal equilibrium during menopause. Harman (2005) and Manson and Bassuk (2011) have delved into the genetic underpinnings of hormone regulation, unraveling its correlation with menopausal symptoms.

Genetic factors extend their influence to vasomotor symptoms, such as hot flashes and night sweats. Avis *et al.* (2001) have probed into the heritability of vasomotor symptoms, suggesting a genetic contribution to their occurrence and severity. Emotional well-being during menopause is also subject to genetic predisposition. Research by Bromberger *et al.* (2003) has explored the genetic foundations of mood swings, depression, and anxiety in the context of the menopausal transition.

The impact of genetic predisposition is further manifested in urogenital changes, influencing symptoms like vaginal dryness and discomfort. Bachmann *et al.* (2002) have investigated the genetic factors influencing physiological alterations in the urinogenital tract during menopause.

Bone health, linked to estrogen decline during menopause, is also subject to genetic influences. The increased risk of osteoporosis associated with this hormonal shift has prompted scientists like Bonaccorsi *et al.* (2006) to delve into the genetic aspects of bone metabolism concerning menopausal bone-related symptoms.

Recognizing the intricate interplay between genetics and menopausal symptoms is paramount for tailoring personalized healthcare approaches. This understanding empowers healthcare professionals to craft interventions based on an individual's genetic makeup, leading to more effective symptom management and enhanced overall well-being during the menopausal transition. While genetics are influential, a comprehensive comprehension of menopausal symptomatology necessitates consideration of the intricate interplay between genetic factors, lifestyle choices, environmental influences, and hormonal dynamics.

Reproductive History

Reproductive history significantly shapes the onset and severity of menopausal symptoms, and various researchers have made valuable contributions to elucidate this intricate relationship. The age at which a woman experiences her first menstrual period, known as menarche, has been identified as a factor influencing menopausal symptoms. Early menarche, associated with a longer reproductive span, may impact the duration and intensity of menopausal symptoms later in life. Santoro *et al.* (2005) conducted research on the age at menarche, exploring its influence on menopausal symptoms, assessing how early or late menarche affects their severity and duration during menopause.

Furthermore, the regularity of menstrual cycles throughout a woman's reproductive years can influence the menopausal transition. Women experiencing irregular menstrual cycles during their reproductive years may undergo a more unpredictable and symptomatic menopausal transition. Mishra *et al.* (2017) delved into this aspect, investigating the association between menstrual irregularities during reproductive years and subsequent menopausal symptoms. Their work shed light on how irregular menstrual cycles might contribute to the variability of menopausal symptoms.

Thomas *et al.* (2006) explored the influence of pregnancy history on menopausal symptoms, examining how the number of pregnancies and childbirths affects hormonal profiles and contributes to the severity of symptoms during the transition to menopause. Their findings suggested that hormonal changes associated with pregnancy and lactation may influence the overall hormonal milieu and subsequent menopausal experience.

Similarly, investigators like McKinlay *et al.* (2005) delved into the role of breastfeeding in modifying hormonal fluctuations and its subsequent impact on menopausal symptoms. Prolonged breastfeeding was found to be associated with the duration and severity of symptoms.

Additionally, Gold *et al.* (2001) investigated the long-term effects of oral contraceptive use on menopausal symptoms. They explored how contraceptive use during reproductive years influences hormonal dynamics and manifests symptoms during menopause.

Mishra *et al.* (2017) also examined the association between the age at menopause and the severity of symptoms, contributing to the understanding of how early or late menopause may impact the overall experience of menopausal symptoms. Collectively, these investigators, among others, have significantly contributed to unraveling the intricate relationship between reproductive history and menopausal symptoms.

Life Style Factors

Lifestyle factors are pivotal in shaping the onset and severity of menopausal symptoms, underscoring the need for a comprehensive healthcare approach during this life stage. Several key elements contribute to the variability and intensity of menopausal symptoms, encompassing physical activity, dietary habits, body weight and composition, smoking and alcohol consumption, stress management, sleep hygiene, and social support.

Research studies, such as those conducted by Abasi *et al.* (2020), have underscored the positive impact of regular physical activity on menopausal symptoms. Exercise has been linked to a reduction in symptom severity, particularly hot flashes, and an enhancement of overall well-being. Investigations by Terachi *et al.* (2010) suggest that dietary choices play a significant role in influencing menopausal symptoms. Diets rich in fruits, vegetables, and whole grains are associated with a lower prevalence and severity of symptoms.

In addition to these lifestyle elements, obesity and a higher body mass index (BMI) have been identified as factors linked to increased severity of menopausal symptoms. Research by Thurston *et al.* (2011) and Huang *et al.* (2018) explores the correlation between body weight and the occurrence of hot flashes and night sweats.

Studies, including those conducted by Gold *et al.* (2000), highlight the potential exacerbating effects of smoking and excessive alcohol consumption on menopausal symptoms. Smoking is associated with an elevated risk of hot flashes, while alcohol intake may contribute to disturbances in sleep patterns. Similarly, high stress levels have been identified as a factor intensifying menopausal symptoms. Investigations by Avis *et al.* (2003) emphasize the importance of stress management techniques, such as mindfulness and relaxation, in mitigating symptom severity.

Sleep disturbances are prevalent during menopause, and researchers like Freeman *et al.* (2015) emphasize the significance of good sleep hygiene practices, including maintaining a consistent sleep schedule and creating a conducive sleep environment, in managing menopausal symptoms.

Social support emerges as a critical factor influencing menopausal experiences. Studies by Bromberger *et al.* (2000) and Avis *et al.* (2001) suggest that women with robust social support networks navigate menopausal symptoms more effectively.

Understanding and addressing these multifaceted lifestyle factors are integral components of a holistic approach to managing menopausal symptoms. Tailoring interventions based on individual lifestyle choices holds the potential to contribute to improved overall well-being during the menopausal transition.

Psychological Factors

Psychological factors play a pivotal role in shaping the onset and severity of menopausal symptoms, contributing significantly to the overall experience of women during this transformative life stage. Recognizing the intricate interplay between the physiological changes linked to menopause and psychological well-being is crucial for adopting a holistic approach to managing menopausal symptoms.

Research studies, exemplified by Avis *et al.* (2003), underscore the substantial impact of stress on menopausal symptoms. Elevated stress levels can intensify manifestations like hot flashes and mood swings. Nevertheless, effective stress management techniques, encompassing mindfulness, relaxation exercises, and coping strategies, have demonstrated their efficacy in alleviating the severity of menopausal symptoms.

Explorations by Joffe and Cohen (1998) delve into the intricate relationship between hormonal fluctuations during menopause and mood swings, anxiety, and depression. Studies, including those conducted by Bromberger *et al.* (2000), propose that women fostering a positive outlook, fortified by robust social support and higher life satisfaction, may encounter menopausal symptoms in a more positive light. Research by Dennerstein *et al.* (1999) suggests that cognitive appraisal, encompassing how women perceive and interpret menopause, can significantly impact symptom severity. Investigations into the role of personality factors in menopausal experiences, while diverse, shed light on the potential influence of traits such as resilience, optimism, and adaptability. Understanding and addressing these psychological factors are vital for tailoring comprehensive interventions that enhance the psychological well-being of women navigating the challenges of menopause.

Ethnic and Cultural Variances

Ethnic and cultural factors can play a significant role in influencing the onset and severity of menopausal symptoms. The experiences and perceptions of menopause can vary across different ethnic and cultural groups, impacting how women navigate this life stage.

Research studies have highlighted the diversity in menopausal experiences among various ethnicities. For example, studies comparing Caucasian, Japanese, Chinese, and African-American women have reported variations in the prevalence and severity of specific symptoms. Gold *et al.* (2006) observed that Caucasian women showed more frequent hot flashes compared to Japanese and Chinese women. Similarly, Japanese and Chinese women experienced more frequent hot flashes compared to African-American women.

Cultural attitudes, beliefs, and practices surrounding menopause can also influence how women experience and cope with symptoms. Sociocultural factors, including societal expectations, family dynamics, and traditional health practices, can shape the perception and management of menopausal symptoms.

Access to healthcare resources and the willingness to seek medical help for menopausal symptoms may differ among ethnic groups, contributing to variations in the reported severity of symptoms. Additionally, cultural norms regarding aging and menopause may influence women's psychological well-being during this transition.

It is essential for healthcare professionals to recognize the impact of ethnic and cultural factors when providing care to women going through menopause. Tailoring interventions to consider cultural sensitivities, beliefs, and practices can contribute to more effective symptom management and a better overall experience for women from diverse backgrounds. Researchers such as Im *et al.* (2010) and Avis *et al.* (2001) have explored the intersection of ethnicity, culture, and menopausal experiences, shedding light on the multifaceted nature of this determinant.

Health Conditions

The presence of health conditions significantly shapes the initiation and severity of menopausal symptoms, with an individual's overall health status playing a pivotal role in molding the menopausal journey. Pre-existing health conditions interact with hormonal changes, contributing to the emergence of symptoms. Women with chronic health issues, including cardiovascular diseases, diabetes, or autoimmune disorders, may encounter more pronounced menopausal symptoms. The interplay of health conditions with hormonal fluctuations impacts the severity of symptoms like hot flashes and mood swings.

Obesity is closely linked to an augmented severity of menopausal symptoms, with a higher body mass index (BMI) correlating with a heightened prevalence of hot flashes and night sweats, as evidenced by research such as that conducted by Thurston *et al.* (2008), exploring the association between body weight and symptom occurrence.

Conditions affecting bone health, such as osteoporosis or osteoarthritis, introduce additional challenges during the menopausal transition. The decline in estrogen, associated with an increased risk of osteoporosis, becomes particularly pertinent for women with compromised bone health.

Women with a history of reproductive health issues, like polycystic ovary syndrome (PCOS) or endometriosis, undergo a distinctive menopausal experience. The interaction between past reproductive health conditions and hormonal changes contributes to the severity of menopausal symptoms.

Pre-existing mental health conditions, such as depression or anxiety, exert influence on the psychological aspects of menopausal symptoms. Women with a history of mental health issues may exhibit greater susceptibility to mood swings, irritability, and alterations in emotional well-being during menopause.

Hence, understanding an individual's health history is imperative for healthcare professionals addressing menopausal symptoms. Tailoring interventions based on the specific health conditions of each woman enhances the effectiveness of symptom management, contributing to an improved overall quality of life during the menopausal transition. Researchers like Freeman *et al.* (2015) and Avis *et al.* (2003) have delved into the intersection of health conditions and menopausal experiences, offering valuable insights into the intricacies of this determinant.

"International Perspective of Menopausal Symptoms Determinants"

Numerous determinants contribute to the manifestation of menopausal symptoms, and among these, racial and ethnic factors emerge as key influencers of both the intensity and extent of these symptoms. A comprehensive study involving 14,906 women from diverse racial, ethnic, and cultural backgrounds in the United States identified racial and ethnic differences as crucial determinants for the prevalence of vasomotor symptoms, including hot flashes and night sweats. Notably, these symptoms were found to be more prevalent in women of Afro-American origin, while women originating from Asian countries experienced them to a lesser extent. Psychosomatic symptoms, on the other hand, were found to be higher in Caucasian women compared to Asian women in the same study (Avis *et al.*, 2001; Khatoon *et al.*, 2018).

Further insights from a Pan-Asia Menopause Group study, encompassing nine Asian ethnic groups, revealed significant variations in the prevalence of menopausal symptoms. Vietnamese women displayed a 100 per cent prevalence of hot flashes, while Indonesian women reported only a 5 per cent prevalence. Muscle and joint pain were prominent vasomotor symptoms experienced by Asian women in general (Haines *et al.*, 2005).

Similar variations were observed in Greek women, where approximately 33 per cent of early post-menopausal individuals exhibited moderate to severe menopausal symptoms, including vasomotor, psychological, and urogenital domains. Age at menopause was identified as a determining factor for the severity of menopausal symptoms in this population (Grigoriou *et al.,* 2013).

Swedish women participating in a separate study demonstrated that poor health conditions and differences in sociodemographic characteristics and lifestyle were governing factors for the manifestation of menopausal symptoms of varying intensities (Li *et al.,* 2003).

In Brazil, a cross-sectional study involving a multi-ethnic group revealed that menopausal women transitioning into their menopausal phase exhibited more vasomotor and urogenital symptoms. Peri-menopausal women also reported memory loss, suggesting a positive correlation with minor psychiatric disorders (Li *et al.,* 2003; Oppermann *et al.,* 2012).

Psychological symptoms linked to low literacy and sociodemographic disparities were reported in a Brazilian study, leading to more non-specific somatic and psychological symptoms. Educational interventions and a focus on the critical window of menopausal transition were recommended (Oppermann *et al.,* 2012).

A Malaysian study with 356 participants aged 40-65 using a modified menopause rating scale revealed that postmenopausal women experienced symptoms like muscle and joint pain, physical and mental exhaustion, anxiety, depressive mood, and irritability. Peri-menopausal and post-menopausal women showed statistically significant differences compared to pre-menopausal women, with post-menopausal women experiencing more urogenital issues (Rahman *et al.,* 2010; Khatoon *et al.,* 2018).

A study in Kawasaki, Japan, conducted by Ishizuka *et al.* (2008), using a self-administered questionnaire on menopause-related symptoms in women aged > 50 years, found a higher incidence of vasomotor and psychological symptoms in post-menopausal women. These symptoms were closely associated with psychosocial and lifestyle factors, as well as body mass index (BMI).

"National Perspective of Menopausal Symptoms Determinants"

In India, the prevalence of menopausal symptoms varies across different states and regions, with factors such as premature menopause (occurring between 35-39 years) exhibiting regional disparities. States like Andhra Pradesh, Gujarat, and Karnataka show varying percentages of premature menopause, with recorded rates of 22.1 per cent, 10.7 per cent, and 10.6 per cent, respectively. Sociodemographic parameters, nutritional factors, fertility-related variables, child-bearing age, and women with no children contribute to this variation. Factors such as low literacy, inadequate diet, low income, and low socioeconomic status are associated with the early onset of menopause (Syamala and Sivakami, 2005).

Furthermore, cultural and traditional influences play a significant role in the manifestation of menopausal symptoms. Menopause introduces several risk factors that can lead to chronic diseases and disability in aging women. The prevalence of menopausal symptoms in urban settings tends to be higher, with premature menopause more likely (Mishra, 2011).

A Pan-India study conducted in 21 cities with 2184 women in their menopause transition or post-menopause found an average menopausal age of 45.59 years. The age of menopause varies between 42 to 49 years, influenced by factors such as nutritional status, medical facilities, educational status, and reproductive factors (Ahuja, 2016).

In the context of Indian culture and tradition, a study in West Bengal assessed the health status of menopausal women and found variations attributed to reproductive history, sociodemographic variables, genetic factors, and lifestyle diseases. Ethnicity was identified as a contributing factor to these variations (Dasgupta *et al.,* 2015).

Studies in different regions of India, including Jammu, Bihar, Agra, Mysore, and Himachal Pradesh, reported varying prevalence and severity of menopausal symptoms. Factors such as age, family composition, menstrual history, personal and family health history, and dietary habits were associated with the manifestation of symptoms (Sharma *et al.,* 2007; Ahsan *et al.,* 2015; Garg *et al.,* 2015; Alva and Chethan, 2016; Kulkarni *et al.,* 2016; Mahajan *et al.,* 2012).

A study in the tribal population of western Odisha identified anxiety, residential status, and education as determining factors for menopausal symptoms (Satpathy and Lenka, 2020).

"Demographic Insights and Menopausal Problems Amongst Tribal Women in Mayurbhanj District, India"

In India, tribal populations constitute 8.6 per cent of the total populace, with significant representation in states like Maharashtra, Odisha, Rajasthan, Chhattisgarh, Andhra Pradesh, Karnataka, Gujarat, Madhya Pradesh, and West Bengal (Census of India, 2001, 2011). These communities, known for their self-reliance, strongly adhere to cultural traditions, often preferring to maintain distinct lifestyles. Mayurbhanj district stands out on the Indian map, boasting lush greenery and a rich cultural heritage. Approximately 56.6 per cent of the district's population comprises various tribal communities, encompassing 30 major tribes and 13 Particularly Vulnerable Tribal Groups (PVTGs) (Sahoo, 2011; Mohanty, 2017).

The Annual Health Survey (2010-11) highlights that 40.7 per cent of women in this region enter marriage before the age of eighteen. The average age at menarche, marking the onset of the menstrual cycle, ranges between 10-16 years (Shah, 1958). Currently, the average menarche age in Odisha is 15.2, and in Mayurbhanj, it is 13.11 years (ICMR, 2003). Various tribal communities exhibit distinct ages at menarche, such as Chhattisgarh tribes with an average of 12.83 years (Verma and Verma, 2014), Delhi tribes at 13.99 years (Dabral and Malik, 2017), Kolkata at 12.3 years, and Maharashtra at 13.38 years (Rokade and Mane, 2009).

Menopause, the final and critical phase of a woman's life, typically occurs between the late 30s and early 50s. Different tribal communities present variations in the mean age of menopause. For instance, Khatri tribal females in Chandigarh experience menopause around 49.5 years, Gujjars of Delhi at 44 years (Dabral and Malik, 2017), rural (urban) Punjabi Brahmin females at 48.22 years (49.30 years), and urban women in Pune, Maharashtra at 45.8 years (Kaulagekar, 2011).

In Mayurbhanj district, the average age at menopause among different tribal women varies: Bhumij - 50.20 years, Bathudi - 42.94 years, and Sabara females - 39 years. These demographic insights shed light on the reproductive health dynamics within the tribal communities of Mayurbhanj district, offering a comprehensive understanding of their unique experiences.

Conclusion

Menopausal symptoms are complex and multifaceted, influenced by various determinants spanning hormonal changes, genetics, reproductive history, lifestyle, psychological factors, cultural variations, and health conditions. Understanding these influences is crucial for healthcare professionals to provide effective and tailored interventions. The Menopause Rating Scale (MRS) emerges as a valuable tool for assessing symptom severity, enabling comprehensive evaluation and monitoring across diverse populations, including the unique context of Mayurbhanj, characterized by its vibrant tribal communities, lush landscapes, and cultural richness. International perspectives highlight significant racial and ethnic variations, emphasizing the need for culturally sensitive approaches. In India, regional, cultural, and socio-economic factors contribute to variations in menopausal experiences. The findings underscore the importance of considering diverse demographic and cultural contexts, infused with the essence of Mayurbhanj, in addressing menopausal symptoms and improving the overall well-being of women during this significant life stage.

REFERENCES

Abasi ELIEH, Keramat AFSANEH and Ghorbani M (2020) The Relationship between Social Support, General Health Status, and Severity of Menopause Symptoms among Postmenopausal Women in Northern Iran. *J. Med. Heal. Sci 14*: 771-776.

Ahsan M, Mallick A, Singh R and Prasad R (2015) Assessment of menopausal symptoms during perimenopause and postmenopause in tertiary care hospital. *J. Basic Clin. Reprod. Sci., 4*(14). doi: 10.4103/2278-960X.153516.

Ahuja M (2016) Age of menopause and determinants of menopause age: A PAN India survey by IMS. *J. Midlife. Health 7*: 126–131. doi: 10.4103/0976-7800.191012.

Ambler DR, Bieber EJ and Diamond MP (2012) Sexual Function in Elderly Women: A Review of Current Literature. *Rev. Obstet. Gynecol. 5:* 16.

Avin Alva B and Chethan T (2016) A study to assess the average age of menopause and menopause-associated symptoms among rural women in Mangalore, Karnataka. *Ntl J Community Med 7*: 404–408.

Avis NE, Stellato R, Crawford S, Bromberger J, Ganz P, Cain V *et al.* (2001) Is there a menopausal syndrome? Menopausal status and symptoms across racial/ethnic groups. *Soc. Sci. Med. 52*: 345–356. doi: 10.1016/S0277-9536(00)00147-7.

Bachmann G, Bancroft J, Braunstein G, Burger H, Davis S, Dennerstein L and Sherwin B (2002) Female androgen insufficiency: the Princeton consensus statement on definition, classification, and assessment. *Fertility and sterility 77*(4): 660-665.

Bagga A (2004) Age and symptomatology of menopause: a case study. *Obs. Gynecol Today 11:* 660–666.

Bagga D, Ashley J, Geffrey S, Wang H, Barnard R, Korenman S *et al.* (1995) Effects of a very low-fat, high-fiber diet on serum hormones and menstrual function. Implications for breast cancer prevention. *Cancer 76:* 2491–2496. doi: https: //doi.org/10.1002/1097-0142.

Belchetz PE, Lloyd MH, Johns RGS and Cohen RD (1973) Effect of late-night calcium supplements on overnight urinary calcium excretion in premenopausal and postmenopausal women. *Br. Med. J. 2:* 510–512.

Bensenor I, Nunes M, Sander Diniz M, Santos I, Brunoni A and Lotufo P (2016) Subclinical thyroid dysfunction and psychiatric disorders: cross-sectional results from the Brazilian Study of Adult Health (ELSA-Brasil). *Clin. Endocrinol. (Oxf) 84:* 250–256. doi: 10.1111/CEN.12719.

Biri A, Bakar C, Maral M, Karabacak O and Bumin MA (2005) Women with and without menopause over the age of 40 in Turkey: Consequences and treatment options. *Maturitas 50:* 167–176.

Bonaccorsi S, Perugini M, Di Bari M, Bellini M, Cervia D, Navarra P and Serio M (2006) Osteocalcin gene HinfI polymorphism and bone mass in postmenopausal Italian women: comparison between Val d'Orcia and Catania. *Bone 38*(2): 173-178.

Booth-LaForce C, Thurston R and Taylor M (2007) A pilot study of a Hatha yoga treatment for menopausal symptoms. *Maturitas 57:* 286–295. doi: 10.1016/J. MATURITAS.2007.01.012.

Borker SA, Venugopalan PP and Bhat SN (2013) Study of menopausal symptoms, and perceptions about menopause among women at a rural community in Kerala. *J. Midlife. Health 4:* 182. doi: 10.4103/0976-7800.118997.

Brambilla D, McKinlay S and Johannes C (1994) Defining the perimenopause for application in epidemiologic investigations. *Am. J. Epidemiol. 140:* 1091–1095.

Bromberger JT, Assmann SF, Avis NE, Schocken M and Kravitz HM (2003) Persistent mood symptoms in a multiethnic community cohort of pre-and perimenopausal women. *American Journal of Epidemiology 158*(4): 347-356.

Bromberger J, Meyer P, Kravitz H, Sommer B, Cordal A, Powell L *et al.* (2001) Psychologic distress and natural menopause: a multiethnic community study. *Am. J. Public Health 91:* 1435–1442. doi: 10.2105/AJPH.91.9.1435.

Castelo-Branco C, Blümel J, Chedraui P, Calle A, Bocanera R, Depiano E, *et al.* (2006) Age at menopause in Latin America. *Menopause 13*: 706–712. doi: 10.1097/01.GME.0000227338.73738.2D.

Cecconi S, Rucci N, Scaldaferri M, Masciulli M, Rossi G, Moretti C, *et al.* (1999) Thyroid hormone effects on mouse oocyte maturation and granulosa cell aromatase activity. *Endocrinology 140:* 1783–1788. doi: 10.1210/ENDO.140.4.6635.

Census of India (2001) Census of India Website/: Office of the Registrar General and Census Commissioner, India. Census of India. Available at: https: //censusindia.gov.in/2011-common/census_data_2001.html [Accessed October 6, 2021].

Census of India (2011). Office of the Registrar General and Census Commissioner, India. Census of India. Available at: https: //censusindia.gov.in/2011-common/censusdata2011.html [Accessed October 6, 2021].

Census of India: Scheduled castes and scheduled tribes (2005) Available at: https: //censusindia.gov.in/census_and_you/scheduled_castes_and_sceduled_tribes.aspx [Accessed September 30, 2021].

Chattha R, Nagarathna R, Padmalatha V and Nagendra H (2008a) Effect of yoga on cognitive functions in climacteric syndrome: a randomized control study. *BJOG 115*: 991–1000. doi: 10.1111/J.1471-0528.2008.01749.X.

Chattha R, Raghuram N, Venkatram P and Hongasandra N (2008b) Treating the climacteric symptoms in Indian women with an integrated approach to yoga therapy: a randomized control study. *Menopause 15:* 862–870. doi: 10.1097/GME.0B013E318167B902.

Chedraui P, Blümel JE, Baron G, Belzares E, Bencosme A, Calle A, *et al.* (2008) Impaired quality of life among middle-aged women: A multicentre Latin American study. *Maturitas 61:* 323–329.

Cheung B, Au T, Chan S, Lam C, Lau S, Lee R, *et al.* (2005) The relationship between hypertension and anxiety or depression in Hong Kong Chinese. *Exp. Clin. Cardiol. 10*: 21.

Chichester VN, Melanie O, Ciranni SC and Patricia J (2011) Approaching Menopause (But Not There Yet!). *Nurs. Womens. Health 15:* 320–364. doi: 10.1111/j.1751-486X.2011.01652.x.

Chim H, Tan BHI, Chia CA, Chew EMD, Yap SC and Seang MS (2002) The prevalence of menopausal symptoms in a community in Singapore. *Maturitas 41:* 275–282. doi: 10.1016/S0378-5122(01)00299-7.

Chobanian A, Bakris G, Black H, Cushman W, Green L, Izzo J, *et al.* (2003) The Seventh Report of the Joint National Committee on Prevention, Detection, Evaluation, and Treatment of High Blood Pressure: the JNC 7 report. *JAMA 289:* 2560–2572. doi: 10.1001/JAMA.289.19.2560.

Chojnacki C, Walecka-kapica E, Klupinska G Pawlowicz M, Blonska A and Chojnacki J (2015) Effects of Fluoxetine and Melatonin on Mood, Sleep Quality And Body Mass Index In Postmenopausal Women. *J. Physiol. Pharmacol. 66:* 665–671.

Chowta N, Sebastian J and Chowta M (2008) Comparative Study Of Menopausal Symptoms In Post-Menopausal And Perimenopausal Women. J. Clin. Diagnostic Res.: 959–962.

Christian D, Kathad M and Bhavsar B (2011) Socio-demographic characteristics of postmenopausal women of the rural area of Vadodara district, Gujarat. *Natl. J. Community Med. 2:* 419–422.

Cohen L, Warneke C, Fouladi R, Rodriguez M and Chaoul-Reich A (2004) Psychological adjustment and sleep quality in a randomized trial of the effects of a Tibetan yoga intervention in patients with lymphoma. *Cancer 100:* 2253–2260. doi: 10.1002/CNCR.20236.

Dabral S and Malik S L (2017) Demographic Study of Gujjars of Delhi II. Reproductive Profile and Mortality Levels. *Kamla Raj Enterp. 16:* 75–82. doi: 10.1080/09709274.2004.11905719.

Dalal PK and Agarwal M (2015) Postmenopausal syndrome. *Indian J. Psychiatry 57:* S222. doi: 10.4103/0019-5545.161483.

Daley A, Stokes-Lampard H and Macarthur C (2009) Exercise to reduce vasomotor and other menopausal symptoms: a review. *Maturitas 63*: 176–180.

Dasgupta D and Ray S (2009) Menopausal Problems Among Rural and Urban Women From Eastern India. *J. Soc. Behav. Heal. Sci. 3*. Available at: https: // scholarworks.waldenu.edu/jsbhs/vol3/iss1/2 [Accessed August 17, 2021].

Dasgupta D, Karar P, Ray S and Ganguly N (2015) Menopausal symptoms and its correlates: A study on tribe and caste population of East India. *Curr. Gerontol. Geriatr. Res. 2015*. doi: 10.1155/2015/984767.

Dennerstein L, Dudley E, Hopper J, Guthrie J and Burger H (2000) A prospective population-based study of menopausal symptoms. *Obstet. Gynecol. 96:* 351–358. doi: 10.1016/S0029-7844(00)00930-3.

Dennerstein L, Lehert P, Burger H and Dudley E (1999) Mood and the menopausal transition. *J. Nerv. Ment. Dis. 187:* 685–691. doi: 10.1097/00005053-199911000-00006.

Dinesh T, Gaur G, Sharma V, Madanmohan T, Kumar KH and Bhavanani A (2015) Comparative effect of 12 weeks of slow and fast pranayama training on pulmonary function in young, healthy volunteers: A randomized controlled trial. *Int. J. Yoga 8:* 22. doi: 10.4103/0973-6131.146051.

Dinnerstein L, Lehert P, Dudley E and Guthrie J (2001) Factors contributing to positive mood during the menopausal transition. *Journal of Nervous and Mental Disorders 189:* 84–89.

Dudley E, Hopper J, Taffe J, Guthrie J, Burger H and Dennerstein L (1998) Using longitudinal data to define the perimenopause by menstrual cycle characteristics. *Climacteric 1:* 18–25. doi: 10.3109/13697139809080677.

Fardella C, Gloger S, Figueroa R, RSantis R, Gajardo C, Salgado C, *et al.* (2000) High prevalence of thyroid abnormalities in a Chilean psychiatric outpatient population. *J. Endocrinol. Invest. 23:* 102–106. doi: 10.1007/BF03343687.

Freedman R (2005) Hot flashes: behavioral treatments, mechanisms, and relation to sleep. *Am. J. Med. 118 Suppl:* 124–130. doi: 10.1016/J.AMJMED.2005.09.046.

Freeman EW and Sherif K (2007) Prevalence of hot flushes and night sweats around the world: a systematic review. *Climacteric 10:* 197–214. doi: 10.1080/13697130601181486.

Freeman EW (2015) Depression in the menopause transition: risks in the changing hormone milieu as observed in the general population. *Women's Midlife Health 1*(1): 1-11.

Fuh J, Wang S, Lu S, Juang K and Chiu L (2001) The Kinmen women-health investigation (KIWI): a menopausal study of a population aged 40-54. *Maturitas 39:* 117–124. doi: 10.1016/S0378-5122(01)00193-1.

Ganapathy T and Furaikh SAl (2018) Health-related quality of life among menopausal women. *Arch. Med. Heal. Sci. 6*: 16–16.

Garg R, Rawat R, Pathak M, Lalitha G and Rajwanshi R (2015) Menopausal Symptoms among Postmenopausal Women of North India: A Cross-sectional Study. *J. South Asian Fed. Menopause Soc. 3:* 3–5. doi: 10.5005/JP-JOURNALS-10032-1060.

Ghianda SD, Tonacchera M and Vitti P (2014) Thyroid and menopause. *Climacteric 17:* 225–234. doi: 10.3109/13697137.2013.838554.

Gold EB, Colvin A, Avis N, Bromberger J, Greendale GA, Powell L, *et al.* (2011) Longitudinal Analysis of the Association Between Vasomotor Symptoms and Race/Ethnicity Across the Menopausal Transition: Study of Women's Health Across the Nation. *96:* 1226–1235.

Gold EB, Crawford SL, Avis NE, Crandall CJ, Matthews KA, Waetjen LE, Lee JS, Thurston R, Vuga M and Harlow SD (2013) Factors Related to Age at Natural Menopause: Longitudinal Analyses From SWAN, *American Journal of Epidemiology, 178 (1)*: 70–83. https: //doi.org/10.1093/aje/kws421

Gold EB, Bromberger J, Crawford S, Samuels S, Greendale G, Harlow S, *et al.* (2001) Factors associated with age at natural menopause in a multiethnic sample of midlife women. *Am. J. Epidemiol. 153:* 865–874.

Gold EB, Colvin A, Avis N, Bromberger J, Greendale G, Powell L, *et al.* (2006) Longitudinal analysis of the association between vasomotor symptoms and race/ethnicity across the menopausal transition: study of women's health across the nation. *Am. J. Public Health 96:* 1226–1235. doi: 10.2105/AJPH.2005.066936.

Gold EB, Sternfeld B, Kelsey J, Brown C, Mouton C, Reame N, *et al.* (2000) Relation of demographic and lifestyle factors to symptoms in a multi-racial/ethnic population of women 40-55 years of age. *Am. J. Epidemiol. 152:* 463–473. doi: 10.1093/AJE/152.5.463.

Grigoriou V, Augoulea A, Armeni E, Rizos D, Alexandrou A, Dendrinos S, *et al.* (2013) Prevalence of vasomotor, psychological, psychosomatic and sexual symptoms in perimenopausal and recently postmenopausal Greek women: association with demographic, life-style and hormonal factors. *Gynecol. Endocrinol. 29:* 125–128.

Guyton AC and Hall JE (2006) Text book of Medical Physiology. 11th ed. *Elsevier Inc.*

Haines C, Xing S, Park K, Holinka C and Ausmanas M (2005) Prevalence of menopausal symptoms in different ethnic groups of Asian women and responsiveness to therapy with three doses of conjugated estrogens/medroxyprogesterone acetate: the Pan-Asia Menopause (PAM) study. *Maturitas 52:* 264–276.

Hardy C, Thorne E and Griffiths A *et al.* (2018) Work outcomes in midlife women: the impact of menopause, work stress and working environment. *Womens Midlife Health 4:* 3. https: //doi.org/10.1186/s40695-018-0036-z

Harlow S, Gass M, Hall J, Lobo R, Maki P, Rebar R, *et al.* (2012) Executive summary of the Stages of Reproductive Aging Workshop + 10: addressing the unfinished agenda of staging reproductive aging. *J. Clin. Endocrinol. Metab. 97*: 1159–1168. doi: 10.1210/JC.2011-3362.

Harman M (2005) Menopause and aging: causes, symptoms, and consequences. *Journal of Gynecological Surgery 21*(4): 139-141.

Hathur B, Basavegowda M and Ashok N (2013) Hypertension: An emerging threat among tribal population of Mysore; Jenu Kuruba tribe diabetes and hypertension study. *Int. J. Heal. Allied Sci. 2:* 270–274.

Heinemann LA, Potthoff P and Schneider HP (2003) International versions of the Menopause Rating Scale (MRS). *Heal. Qual. Life Outcomes 11(1)*: 1–4. doi: 10.1186/1477-7525-1-28.

Higginbotham S (2009) Contraceptive considerations in obese women: release date 1 September 2009, SFP Guideline 20091. *Contraception 80:* 583–590.

Hunter M (1996) Depression and the menopause. *BMJ 313:* 1217–1218. doi: 10.1136/BMJ.313.7067.1217.

Hunter M and Rendall M (2007) Bio-psycho-socio-cultural perspectives on menopause. *Best Pract. Res. Clin. Obstet. Gynaecol. 21:* 261–274. doi: 10.1016/J.BPOBGYN.2006.11.001.

ICMR (2003) Health Status of Primitive Tribes of Orissa. Icmr Bull. Retrieved from https: //main.icmr.nic.in/sites/default/files/icmr_bulletins/BUOCT03.pdf

IIPS and ORC (2007) National Family Health Survey (NFHS-3). Mumbai: IIPS. Retrieved from https: //dhsprogram.com/pubs/pdf/frind3/frind3-vol1andvol2.pdf

Im EO, Lee B, Chee W, Dormire S and Brown A (2010) A national multiethnic online forum study on menopausal symptom experience. *Nursing Research 59*(1): 26-33.

Innes KE, Selfe T and Taylor A (2008) Menopause, the metabolic syndrome, and mind-body therapies. Menopause 15: 1005. https: //doi.org/10.1097/01.GME.0B013E318166904E

Iosif C and Bekassy Z (1984) Prevalence of genito-urinary symptoms in the late menopause. Acta Obs. Gynecol Scand 63: 257–260.

Ishizuka B, Kudo Y and Tango T (2008) Cross-sectional community survey of menopause symptoms among Japanese women. Maturitas 61: 260–267.

James W, Ferro-Luzzi A and Waterlow J (1988) Definition of chronic energy deficiency in adults. Report of a working party of the International Dietary Energy Consultative Group. Eur. J. Clin. Nutr. 42: 969–981. Retrieved from https: //europepmc.org/article/med/3148462

Jaszmann LT, Lith ND and Zaat JC (1969) The age at menopause in the Netherlands: The statistical analysis of a survey. Int. J. Fertil. 14: 106–117.

Jelliffe DB (1966) The assessment of the nutritional status of the community (with special reference to field surveys in developing regions of the world). Monogr Ser World Heal. Organ 53: 3–271.

Joffe H and Cohen L (1998) Estrogen, serotonin, and mood disturbance: where is the therapeutic bridge? Biol. Psychiatry 44: 798–811. https: //doi.org/10.1016/S0006-3223(98)00169-3

Joffe H and Lee SC (1998) Estrogen, serotonin, and mood disturbance: where is the therapeutic bridge?. *Biological psychiatry 44* (9): 798-811.

Kalarhoudi MA, Taebi M, Sadat Zand Saberi F (2011) Assessment of Quality of Life in Menopausal Periods: A Population Study in Kashan, Iran. Iran. Red Crescent Med. J. 13: 811.

Kalogeraki A, Tamiolakis D, Relakis K, Karvelas K, Froudarakis G, Hassan E, *et al.* (1996) Cigarette smoking and vaginal atrophy in postmenopausal women. In Vivo (Brooklyn) 10: 597–600.

Kapur P, Sinha B and Pereira BMJ (2009) Measuring climacteric symptoms and age at natural menopause in an Indian population using the Greene Climacteric Scale. Menopause 16: 378–384. https: //doi.org/10.1097/GME.0B013E31818A2BE9

Karmakar N, Majumdar S, Dasgupta A and Das S (2017) Quality of life among menopausal women: A community-based study in a rural area of West Bengal. J. Midlife. Health 8: 21. https: //doi.org/10.4103/JMH.JMH_78_16

Kato I, Toniolo P, Akhmedkhanov A, Koenig K, Shore R and Zeleniuch-Jacquotte A (2010) Prospective study of factors influencing the onset of natural menopause. J. Clin. Epidemiol. 51: 1271–1276.

Kaufert P and Syrotuik J (1981) Symptom reporting at the menopause. Soc. Sci. Med. 15: 173–184.

Kaufert P, Gilbert P and Tate R (1992) The Manitoba Project: a re-examination of the link between menopause and depression. Maturitas 14: 143–155. https: //doi.org/10.1016/0378-5122(92)90006-P

Kaulagekar A (2011) Age of menopause and menopausal symptoms among urban women in Pune, Maharashtra. J. Obstet. Gynecol. India 61: 323–326. https: // doi.org/10.1007/S13224-011-0036-5

Kaur M and Talwar I (2009) Age at natural menopause among rural and urban Punjabi Brahmin females. Anthropologist 11: 255–258. https: //doi.org/10. 1080/09720073.2009.11891112

Kessenich CP and Cathy AR (2013) Inevitable Menopause. Obstet. Gynecol. 56: 155–172.

Khalsa S (2004) Treatment of chronic insomnia with yoga: a preliminary study with sleep-wake diaries. Appl. Psychophysiol. Biofeedback 29: 269–278. https: // doi.org/10.1007/S10484-004-0387-0

Khatoon F, Sinha P, Shahid S and Gupta U (2018) Assessment of menopausal symptoms using modified menopause rating scale (MRS) in women of Northern India. Int. J. Reprod. Contraception, Obstet. Gynecol. 7: 947–951. https: //doi. org/10.18203/2320-1770.IJRCOG20180871

Krassas GE (2000) Thyroid disease and female reproduction. Fertil. Steril. 74: 1063–1070. https: //doi.org/10.1016/S0015-0282(00)01589-2

Kriplani A and Banerjee K (2005) An overview of age of onset of menopause in northern India. Maturitas 52: 199–204.

Kronenberg F (1990) Hot Flashes: Epidemiology and Physiologya. Ann. N. Y. Acad. Sci.592: 52–86. https: //doi.org/10.1111/J.1749-6632.1990.TB30316.X

Kshatriya G and Acharya S (2016) Triple Burden of Obesity, Undernutrition, and Cardiovascular Disease Risk among Indian Tribes. PLoS One 11. https: //doi. org/10.1371/JOURNAL.PONE.0147934

Ku SY, Kang JW, Kim H, Ku PS, Lee SH, Sub CS, *et al.* (2004) Regional differences in age at menopause between Korean - Korean and Korean - Chinese. Menopause 11: 569–574.

Kulkarni P, Savita R, Kumar D and Manjunath R (2016) Burgeoning menopausal symptoms: An urgent public health concern. J. Midlife. Health 7: 83. https: //doi.org/10.4103/0976-7800.185329

Lawlor DA, Ebrahim S and Smith GD (2003) The association of socio-economic position across the life course and age at menopause: the British Women's Heart and Health Study. *BJOG An Int. J. Obstet. Gynaecol. 110*(9): 1078–1087. https: //doi.org/10.1111/J.1471-0528.2003.02519.X.

Lenka C (2016) Knowledge, practice and lifestyle management of postmenopausal women: A study in Odisha. *Int. J. Home Sci. 2:* 80–84. Retrieved from www.homesciencejournal.com [Accessed August 31, 2021].

Li C, Samsioe G, Borgfeldt C, Lidfeldt J, Agardh C and Nerbrand C (2003) Menopause-related symptoms: What are the background factors? A prospective population-based cohort study of Swedish women (The Women's Health in Lund Area study). *Am. J. Obstet. Gynecol. 189*(6): 1646–1653. https: //doi.org/10.1016/S0002-9378(03)00872-X.

Li J (2006) Inflammation in hypertension: Primary evidence. *Chin. Med. J. (Engl) 119*(14): 1215–1221.

Llaneza P, García-Portilla MP, Llaneza-Suárez D, Armott B and Pérez-López FR (2012) Depressive disorders and the menopause transition. *Maturitas 71:* 120–130.

Luoto R, Kaprio J and Uutela A (1994) Age at natural menopause and sociodemographic status in Finland. *Am. J. Epidemiol. 139*(1): 64–76.

Magurský V, Mesko M and Sokolík L (1975) Age at the menopause and onset of the climacteric in women of Martin District, Czechoslovakia: Statistical survey and some biological and social correlations. *Int. J. Fertil. 20:* 17–23.

Mahajan N, Aggarwal M and Bagga A (2012) Health issues of menopausal women in North India. *J. Midlife. Health 3:* 84. https: //doi.org/10.4103/0976-7800.104467.

Maier W, Buller R, Philipp M and Heuser I (1988) The Hamilton Anxiety Scale: Reliability, validity and sensitivity to change in anxiety and depressive disorders. *J. Affect. Disord. 14*(1): 61–68. https: //doi.org/10.1016/0165-0327(88)90072-9.

Maki PM, Rubin LH, Fornelli D and Drogos L (2008) Banishing the blues: hotter flashes, flashier, and sleepier? Increased insomnia symptoms in women with a history of major depression. *Menopause 15*(2): 252-261.

Manimunda SP, Sugunan AP, Benegal V, Balakrishna N, Rao MV and Pesala KS (2011) Association of hypertension with risk factors and hypertension-related behaviour among the aboriginal Nicobarese tribe living in Car Nicobar Island, India. *Indian J. Med. Res. 133*: 287.

Manson JE and Bassuk SS (2011) Invited commentary: Hormone therapy and risk of coronary heart disease. *BMJ 342:* d1401.

Mariotti S, Franceschi C, Cossarizza A and Pinchera A (1995) The aging thyroid. *Endocr. Rev. 16*(6): 686–715. https: //doi.org/10.1210/EDRV-16-6-686.

Marshall FH (1910) *The Physiology of Reproduction.* Longmans: London.

Matthews KA, Wing RR, Kuller LH, Meilahn EN, Kelsey SF, Costello EJ, *et al.* (1990) Influences of natural menopause on psychological characteristics and symptoms of middle-aged health women. *J. Consult. Clin. Psychol. 58*(3): 345–351.

McKinlay JB, McKinlay SM and Brambilla D (*1987)* The Relative Contributions of Endocrine Changes and Social Circumstances to Depression in Mid- Aged Women. *Journal of Health and Social Behavior 28*(4): 345-63.

McKinlay SM, Bifano NL and McKinlay JB (1985) Smoking and age at menopause in women. *Annab Intern. Med. 103*(2): 305–306.

Meeta DL, Agarwal N, Vaze N, Shah R and Malik S (2013) Clinical practice guidelines on menopause: An executive summary and recommendations. *J. Midlife. Health 4:* 77. https: //doi.org/10.4103/0976-7800.115290.

Mehta K, Imonsick E, Penninx B, Schulz R, Rubin S Satterfield S, *et al.* (2003) Prevalence and correlates of anxiety symptoms in well-functioning older adults: Findings from the health aging and body composition study. *J. Am. Geriatr. Soc. 51*(4): 499–504. https: //doi.org/10.1046/J.1532-5415.2003.51158.X.

Melby M, Lock M and Kaufert P(2005) Culture and symptom reporting at menopause. *Hum. Reprod. Update 11*(5): 495–512.

Meldrum D, Defazio J, Erlik Y Lu J, Wolfsen A, Carlson H, *et al.* (1984) Pituitary hormones during the menopausal hot flash. *Obstet. Gynecol. 64*(5): 752–756.

Meschia M, Pansini F, Modena AB, Aloysio D, Gambacciani M, Parazzini F, Campagnoli C, *et al.* (2000) Determinants of age at menopause in Italy: Results from a large cross-sectional study. ICARUS Study Group. Italian Climacteric Research Group Study. *Maturitas 34*(2): 119–125.

Meshram I, Arlappa N, Balkrishna N, Rao K, Laxmaiah A and Brahmam G (2012) Prevalence of hypertension, its correlates and awareness among adult tribal population of Kerala state, India. *J. Postgrad. Med. 58:* 255. https: //doi. org/10.4103/0022-3859.105444.

Milsom I and Molander U (1998) Urogenital ageing. *J. Br. Menopause Soc. 4*(4): 151–156.

Mishra GD, Pandeya N, Dobson AJ, Chung HF, Anderson D, Kuh D, Sandin S, Giles GG, Bruinsma F, Hayashi K, Lee JS, Mizunuma H, Cade JE, Burley V, Greenwood DC, Goodman A, Simonsen MK, Adami HO, Demakakos P and Weiderpass E. (*2017)* Early menarche, nulliparity and the risk for premature and early natural menopause. *Hum Reprod. 32*(3): 679-686. doi: 10.1093/humrep/dew350. PMID: 28119483; PMCID: PMC5850221.

Mishra G, Brown W and Dobson A (2003) Physical and mental health: Changes during menopause transition. *Qual. Life Res. 12*(4): 405–412. https: //doi.org/10.1023/A: 1023421128141.

Mishra S (2011) Menopausal transition and postmenopausal health problems: A review on its bio-cultural perspectives. *Health (Irvine. Calif). 3:* 233–237. https: //doi.org/10.4236/HEALTH.2011.34041.

Mohammad K, Hashemi S and Farahani F (2004) Age at natural menopause in Iran. *Maturitas 49:* 321–326.

Mohanty B (2017) Tribal population of Mayurbhanj. *ijiras.com, 4*. Retrieved from http: //www.ijiras.com/2017/Vol_4-Issue_1/paper_17.pdf [Accessed August 17, 2021].

Mohapatra S, Satapathy K, Patra P and Das P (2015) Epidemiology of blood pressure among some tribes of India: Bhumij, Bathudi and Savar. *Tribal Trib. 7:* 1–4.

Nagar S and Dave P(2017). Perception of Women Towards Physiological Problems Faced at Menopause. *Kamla Raj Enterp. 7:* 173–175. https: //doi.org/10.1080 /09720073.2005.11890902.

Nappi R and Kokot-Kierepa M (2012) Vaginal Health: Insights, Views and Attitudes (VIVA) - Results from an international survey. *Climacteric 15*(1) 36–44. https: //doi.org/10.3109/13697137.2011.647840.

Neugarten BL and Kraines RJ (1965) Menopausal symptoms' in women of various ages. *Psychosom. Med. 27*(3): 266–276.

Nicol-Smith L (1996) Causality, menopause, and depression: A critical review of the literature. *BMJ 313*: 1229–1232. https: //doi.org/10.1136/BMJ.313.7067.1229.

Oppermann K, Fuchs S, Donato G, Bastos C and Spritzer P (2012) Physical, psychological, and menopause-related symptoms and minor psychiatric disorders in a community-based sample of Brazilian premenopausal, perimenopausal, and postmenopausal women. *Menopause 19*(3): 355–360. doi: 10.1097/GME.0B013E31822BA026.

Palacios S, Henderson VW, Siseles N, Tan D and Villaseca P (2010) Age of menopause and impact of climacteric symptoms by geographical region. *Gynecological Endocrinology 26*(6): 419–428. doi: 10.3109/13697137.2010.507886.

Pastore L, Carter R, Hulka B and Wells E (2004) Self-reported urogenital symptoms in postmenopausal women: Women's Health Initiative. *Maturitas 49*(3): 292–303.

Paterniti S, Alpérovitch A, Ducimetière P, Dealberto M, Lépine J and Bisserbe J (1999) Anxiety but not depression is associated with elevated blood pressure in a community group of French elderly. *Psychosomatic Medicine 61*(1): 77–83. doi: 10.1097/00006842-199901000-00013.

Peeyananjarassri DK, Cheewadhanaraks S, Hubbard M, Manga RZ, Manocha R and Eden J. (2009) Menopausal symptoms in a hospital-based sample of women in southern Thailand. *Climacteric 9*(1): 23–29. doi: 10.1080/13697130500487422.

Poomalar G and Arounassalame B (2013) The Quality of Life During and After Menopause Among Rural Women. *Journal of Clinical and Diagnostic Research 7*. doi: 10.7860/JCDR/2012/4910.2688.

Prior J (2013) Perimenopause. *Central Menstrual Cycle and Ovulation Research (CeMCOR), Asian Journal of Medical Sciences 3:* 154–157.

Rahman SASA, Zainudin SR and Mun VLK(2010) Assessment of menopausal symptoms using modified Menopause Rating Scale (MRS) among middle age women in Kuching, Sarawak, Malaysia. *Asia Pacific Family Medicine 9*(1): 1–6. doi: 10.1186/1447

Reed SD, Newton KM, LaCroix AZ, Grothaus LC, Grieco VS, Ehrlich K and Carpenter JS (2008) Vaginal, endometrial, and reproductive hormone findings: randomized, placebo-controlled trial of black cohosh, multibotanical herbs, and dietary soy for vasomotor symptoms: the Herbal Alternatives for Menopause (HALT) Study. *Menopause 15*(1): 51-58.

Santoro N, Brown JR, Adel T and Skurnick JH (1999) Characterization of reproductive hormonal dynamics in the perimenopause. *The Journal of Clinical Endocrinology and Metabolism 84*(3): 846-851.

Tepper R, Neri A, Kaufman H, Schoenfeld A and Ovadia J (1987) Menopausal hot flushes and plasma Beta-endorphins. *Obstetrics and Gynecology 70*(1): 150–152.

Terachi M, Obayashi S, Akiyoshi M, Kato K, Matsushima E and Kubota T (2010) Insomnia in Japanese Peri- and Postmenopausal Women. *Climacteric 13:* 479–496.

Thomas SL and Ellertson C (2000) Nuisance or natural and healthy: should monthly menstruation be optional for women?. *The Lancet 355* (9207): 922-924.

Thompson B, Hart SA and Dumo D (1973) Menopausal age and symptomatology in a general practice. *Journal of Biosocial Science 5*(1): 71–82.

Thurston R and Joffe H (2011) Vasomotor symptoms and menopause: findings from the Study of Women's Health across the Nation. *Obstetric and Gynecologic Clinics of North America 38*(3): 489–501. https: //doi.org/10.1016/J.OGC.2011.05.006

Torgerson D, Avenell A, Russell I and Reid D (1994) Factors associated with onset of menopause in women aged 45-49. *Maturitas 19*(2): 83–92. https: //doi.org/10.1016/0378-5122(94)90057-4

Unni (2010) Third consensus meeting of Indian Menopause Society (2008): A summary. *Journal of Mid-life Health 1*(2): 43. https: //doi.org/10.4103/0976-7800.66987

Utian WH (2005)Psychosocial and socioeconomic burden of vasomotor symptoms in menopause: A comprehensive review. *Health and Quality of Life Outcomes 3*(1) 47. https: //doi.org/10.1186/1477-7525-3-47

Vaze N and Joshi S (2010)Yoga and menopausal transition. *Journal of Mid-life Health 1*(2): 56. https: //doi.org/10.4103/0976-7800.76212

Ven van de A, Muntjewerff J, Netea-Maier R, de Vegt F, Ross H, Sweep F, *et al.* (2012) Association between thyroid function, thyroid autoimmunity, and state and trait factors of depression. *Acta Psychiatrica Scandinavica 126*(5) 377–384. https: //doi.org/10.1111/J.1600-0447.2012.01870.X

Verma DK and Verma O (2014) Reproductive Life of the Bhatra Women of Bastar, Chhattisgarh. *International Journal of Humanities, Social Sciences and Education 1:* 2349. Retrieved from www.arcjournals.org [Accessed October 6, 2021].

Warren MP (2009) Menopausal hot flashes: How long do they really last? *Menopause 16*(2): 427. https: //doi.org/10.1097/GME.0B013E3181A1BF1D

Whiteman MK, Staropoli CA, Langenberg PW, McCarter RJ, Kjerulff KH and Flaws JA (2003) Smoking, body mass, and hot flashes in midlife women. *Obstetrics and Gynecology 101*(2): 264–272. https: //doi.org/10.1016/S0029-7844(02)02593-0

Wilbush J (1979) Lza Menespausle: The birth of a syndrome. *Maturitas 1*(3): 145–151.

Winokur G (1973) Depression in the menopause. *American Journal of Psychiatry 1*(8): 92–93.

Woods N, Mariella A and Mitchell E (2006) Depressed mood symptoms during the menopausal transition: observations from the Seattle Midlife Women's Health Study. *Climacteric 9*(3): 195–203.

World Health Organization Expert Consultation (2004) Appropriate body-mass index for Asian populations and its implications for policy and intervention strategies. *The Lancet 363*(9403): 157–163. https: //doi.org/10.1016/S0140-6736(03)15268-3

World Health Organization (1996a) Fighting disease, fostering development/ report of the Director-General. Available at: https: //apps.who.int/iris/ handle/10665/36848

World Health Organization (1996b) Research on the menopause in the 1990s. Available at: https: //ci.nii.ac.jp/naid/10008975956

Wright AL (1981) On the calculation of climacteric symptoms. *Maturitas 3*(1): 55–63.

Wu X, Huang J, Shen C, Liu Y, He S, Sun J, *et al.* (2020) NRF2 deficiency increases obesity susceptibility in a mouse menopausal model. *PLoS ONE 15*(2): e0228559. https: //doi.org/10.1371/ journal.pone.0228559

Yadlapalli K, Babu B and Naidu J (2004) Prevalence of hypertension in some cross-cultural populations of Visakhapatnam District, South India. *Ethnicity and Disease 14*(2): 250–259.

Chapter 9

Toxic Effect of Heavy Metals (Lead and Cadmium) on Gill of Fish

Puspanjali Parida and Monalisha Nayak

P.G. Department of Zoology, Maharaja Sriram Chandra Bhanja Deo University, Baripada–757003, Odisha
e-mail;: p_parida2000@yahoo.co.in

Introduction

Heavy metals constitute a major source of water pollutant. Heavy metals enter into the fish body by three possible ways (a) through gill (b) by digestive track (c) absorption through body surface. Among these, gills are considered as a significant site for direct uptake of heavy metals from fresh water (Romeo *et al.*, 1999; Beijer*e al.*, 1986). Heavy metals accumulate in the tissues of fish and other aquatic organisms and passes through food chain causes biomagnifications of heavy metals at different trophic level. Alteration in the chemical composition of a natural aquatic environment by hazardous substances like heavy metals, Pesticides and industrial effluents usually affect the behavioural, biochemical and physiological parameters of aquatic fauna including fishes (Radhaia *et al.*, 1987).

In the present work, biochemical parameter of gill of *Catla catla* were measured by exposing the fishes with cadmium chloride and lead nitrate at different time interval (24, 48, 72 hour) in laboratory conditions.

Materials and Methods

Animal

Fishes (*Catla catla*) of 80-90 days, length: 9-12 cm and weight: 9-12 g were procured from local fish farm, Baisingha, Mayurbhanj, Odisha and transported in polythene bag half filled with pond water and oxygen from oxygen cylinder. They were acclimatized for seven days in laboratory condition before experiment in an aerated condition. These fishes were fed with fish feed. Day and night cycle were maintained. Tetracycline hydrochloride (water soluble powder) was used as a disinfectant.

Experimental Set Up

Four number of aquarium (50 cm x 40 cm x 50 cm) filled with 20 litre tap water, treated with Tetracycline hydrochloride powder were prepared. Healthy fishes were segregated in two groups: - (i) Control (C_{Cd}/C_{Pb}) and (ii) Experimental (Subgroups: $E1_{Cd}/E1_{Pb}$, $E2_{Cd}/E2_{Pb}$ and $E3_{Cd}/E3_{Pb}$).

Aquarium $CdCl_2/Pb(NO_3)_2$	*Quantity $CdCl_2/Pb(NO_3)_2$ in mg/L*	*Time Interval (Hour)*
C_{Cd}/C_{Pb}	—	0, 24, 48, 72
$E1_{Cd}/E1_{Pb}$	0.453/2.685	24, 48, 72
$E2_{Cd}/E2_{Pb}$	0.566/3.356	24
$E3_{Cd}/E3_{Pb}$	0.906/5.370	24

Preparation of Supernatant

Catla catla, 5 in number were picked up from the desired aquarium and their weight was measured separately in monopan balance (Shimadzu; ELB 300). A 10 per cent homogenate of gill tissue was prepared in ice-cold 50mM phosphate buffer (pH 7.4) using pre-chilled porcelain motor and pestle by up and down strokes at 4°C. The homogenate was centrifuged and the supernatant was taken for biochemical assay.

Protein Estimation

Protein estimation of the samples was made according to the method of Lowry *et al.* (1951). To 0.1 ml of 10 per cent homogenate, 0.4 ml of distilled water, 5 ml of biuret reagent containing 2 per cent Na_2Co_3 (Sodium Carbonate) in 0.1N NaOH, 1 per cent $KNaC_4H_6O_6$ (Potassium-sodium tartarate tetrahydrate) and 0.5 per cent $CuSO_4$ (Copper II Sulphate pentahydrate) in the ratio 100: 2: 2 was added and vortexed. After 10 minutes of incubation, 0.5 ml of Folin-ciocalteau's reagent (1 part of commercially available reagent diluted with 2 parts of distilled water) was added and vortexed. After 30 minutes of incubation at room temperature absorbance was measured at 700 nm by spectrophotometer (Systronics Visiscan: 167), against an appropriate blank. Protein content was expressed as mg/g weight of the tissue and aqueous BSA (Bovine serum albumin) was taken as standard protein.

Lipid Peroxidation Estimation

Lipid Peroxidation assay was determined by Thiobarbituric acid test (TBA test) according to Okhawa *et al*, 1979. 1.9 ml. of TBA reagent containing 8.1 per cent (w/v) Sodium dodecyl sulphate (SDS), 20 per cent (V/V) acetic acid pH 3.5, 0.5 ml. of 0.8 per cent (w/v) aqueous solution of Butylated hydrotoluene (BHT), Thiobarbituric acid (TBA) and distilled water was added to 0.1ml. of the supernatant and mixed thoroughly. Test substance was then boiled in a water bath for 1 hour, using glass balls as condenses. Then the test tubes were cooled to room temperature and centrifuged at 4500 rpm (1000×g) for 10 minutes at room temperature in Table Top Laboratory Centrifuge (Remi). Absorbance of supernatant was measured at 532nm. The concentration of TBARS was calculated from extinction coefficient of 156 mM^{î1} cm^{î1} (Wills, 1969) and expressed as n mole equivalent of MDA formed per mg protein.

Reduced Glutathione Estimation

GSH content of the tissue samples was determined according to the method of Ellman, (1959) with slight modification. 0.7ml. of the supernatant was precipitated in ice cold Tri-carboxylic acid (10 per cent concentration) and centrifuged at 4500 rpm (1000×g) for 10 minute at 4°C and the supernatant was used for GSH assay. 0.5ml. supernatant was mixed with 2.5 ml of DTNB (5, 5-Dithiobis 2 Nitro benzoic acid) and incubated at room temperature. Absorbance at 412 nm was recorded against a blank containing only DTNB within 5-30 minutes. The GSH content of the tissue was expressed as mg/g tissue.

Statistical Methods

One-way ANOVA and Post Hoc analysis was carried out to find out the level of significance between *Catla catla* treated with $CdCl_2$ and $Pb(NO_3)_2$ over a period of 24 hour, 48 hour, 72 hour and in control. A difference was taken as significant when P was less than 0.05. Statistics is done with the help of software SPSS package 23.0.

Results and Discussion

Toxicity of Cadmium Chloride ($CdCl_2$) and Lead Nitrate [$Pb(NO_3)_2$] on Morphology and Behaviour of *Catla catla*

The toxic effect of Cadmium Chloride and Lead Nitrate on *Catla catla* was recorded at 0 hour (control), 24 hour, 48 hour and 72 hour along with different concentration at 24 hour. Present study reveals that the fishes showed normal behaviour such as well coordinated active movement,normal gill movement, free gulping, horizontal hanging in water and natural body colour before exposed heavy metals. After exposed to heavy metals number of morphological and behavioural changes like fishes frequently came to water surface,tried to jump out of water, fin degradation,haemorrhage on skin, fading of their body colour, restlessness and excess mucus secretion from body were observed. Slight behavioural variations were observed between fishes exposed to two different heavy metals. During Lead Nitrate exposure fishes became more restless, showed fin degradation and secret more mucus in comparison to Cadmium Chloride exposure. During cadmium chloride exposure haemorrhages were found at head and ventral region of the body.

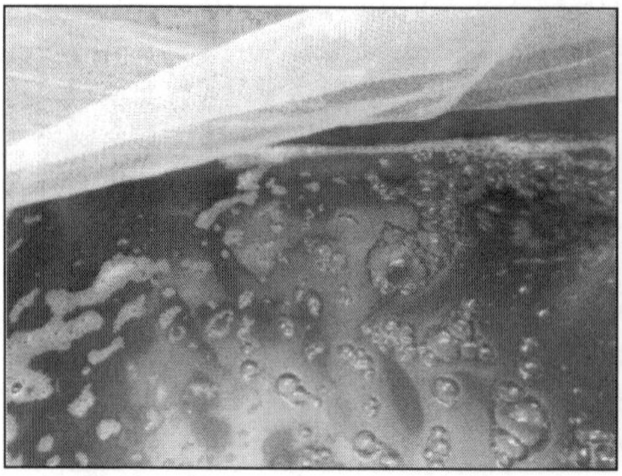

Figure 9.1 Normal Behaviour of *Catla catla* in Aquarium (Foam formation due to tetracycline).

Figure 9.2: Behavioural Changes in the Heavy Metal Exposed Fish (Fishes come frequently to the surface of water).

Figure 9.3: Unexposed Fish

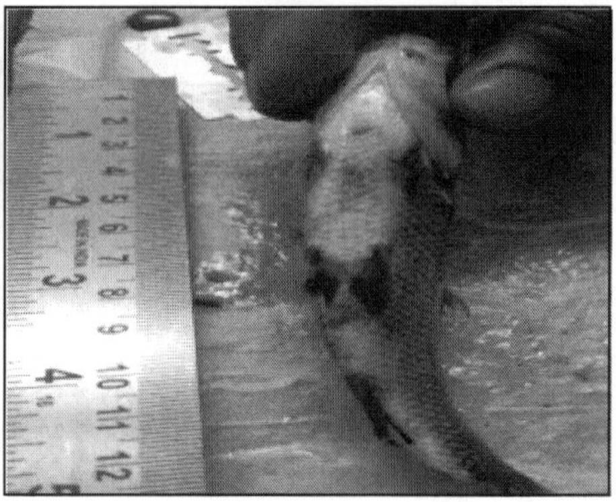

Figure 9.4. Appearance of Haemorrhage in Ventral Part of the Body after exposed to $CdCl_2$

Protein Content

i) Cadmium Chloride

Protein content (mg/g tissue) in gill of *Catla catla* were 22.44±1.768, 18.800±0.881, 17.55±0.363, 19.23±1.601 at 0-hour, 24 hour, 48 hour and 72 hour respectively (Figure 9.9). In other words, protein content decreased at

24, 48 and 72 hour in comparison to control group and was lowest at 48 hour. One way ANOVA revealed that the protein content at different time interval was significantly different [$F(3,19)=10.418, P=0.000$]. Post Hoc analysis revealed that mean difference of protein content at different time interval when exposed with cadmium Chloride in gill of *Catla catla* was significant at control, 24 hour,48 hour and 72 hour($P<0.05$;LSD).

Figure 9.5: Head Region of Normal Fish.

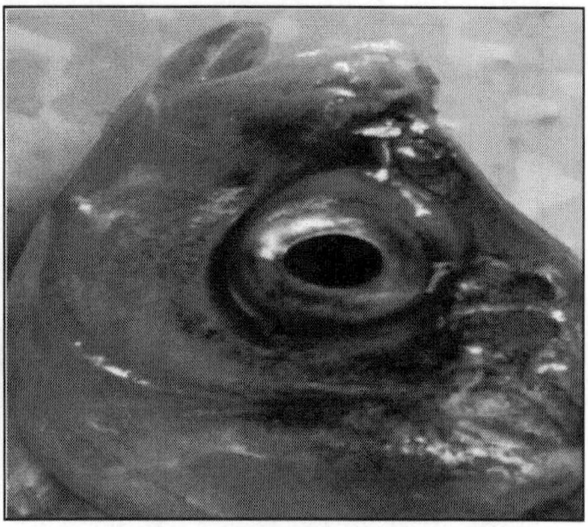

Figure 9.6: Appearance of Haemorrhage Near Mouth after Exposed to Cadmium.

Figure 9.7: Fin Rays of Unexposed Fish.

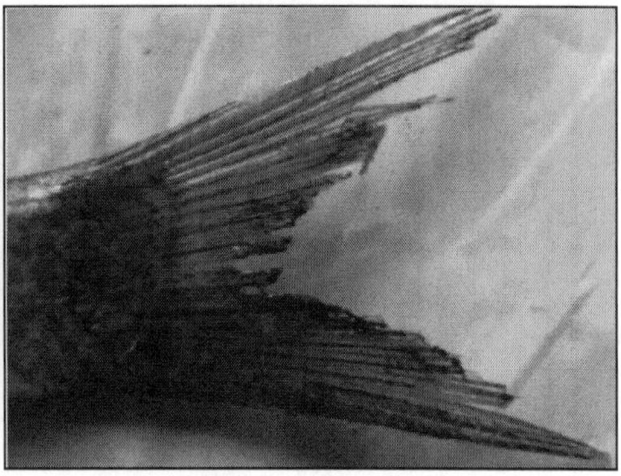

Figure 9.8: Damage of Fin Rays after Exposed Lead.

ii) Lead Nitrate

Protein content (mg/g tissue) in gill of *Catla catla* was 21.489±0.861 in control and after exposed with Lead Nitrate protein content were 21.346±3.184, 17.562±2.605, 12.882±2.121, at 24-hour, 48 hour and 72 hour respectively (Figure 9.10). One way ANOVA revealed that the protein content at different time interval is significant [$F(3,19)=14.868, P=0.000$]. Post Hoc analysis revealed that mean difference of protein content at different time interval when exposed with lead nitrate in gill of *Catla catla* decreased significantly at 0-hour, 24-hour,48 hour and 72 hours (P <0.05;LSD)

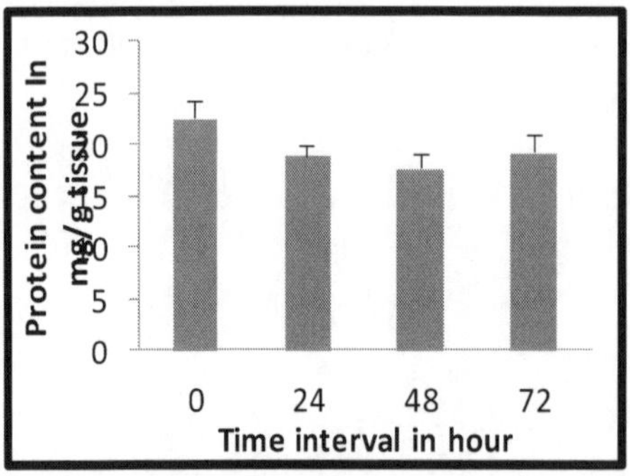

Figure 9.9: Comparison of Protein Content in *Catla catla* Exposed to $CdCl_2$ (0.453mg/L) different at Time Interval.

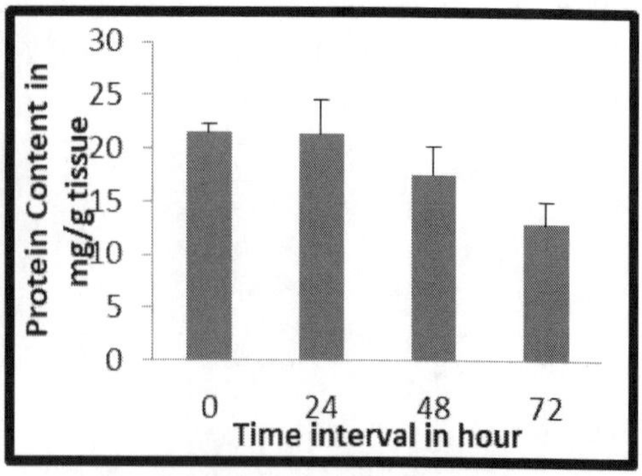

Figure 9.10: Comparison of Protein Content in *Catla catla* Exposed to $Pb(NO_3)_2$ (2.658mg/L) different at Time Interval.

Lipid Peroxidation (LPx)

i) Cadmium Chloride

LPx (n mole of TBARS/mg protein) of *Catla catla* is 5.672 ± 0.445 in unexposed and after exposed to Cadmium Chloride LPx content were 6.0452 ± 0.905 6.258± 0.336, 5.428 ±0.423 at 24 hour,48 hour and 72hour respectively (Figure 9.11). One way ANOVA revealed that the LPx level at different time interval differed significantly [F(3,19) =3.013,P=0.061]. Post Hoc analysis revealed that mean difference LPx in gill of *catla catla*, when exposed with cadmium Chloride was significantly different in control, 24 hour,48 hour and 72 hour (P<0.05; LSD).

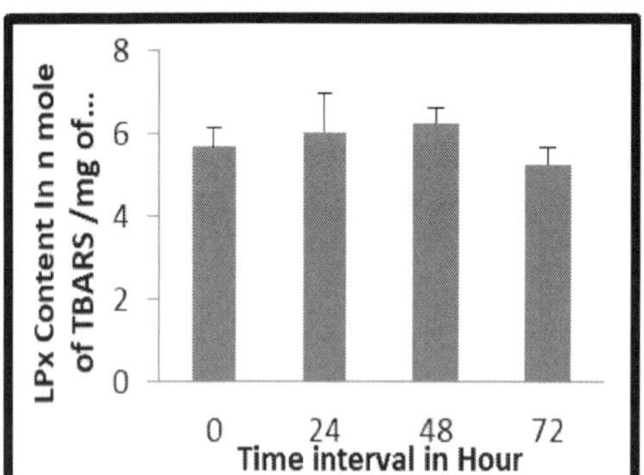

Figure 9.11: Comparison of LPx in *Catla catla* Exposed to $CdCl_2$ (0.453mg/L) different at Time Interval.

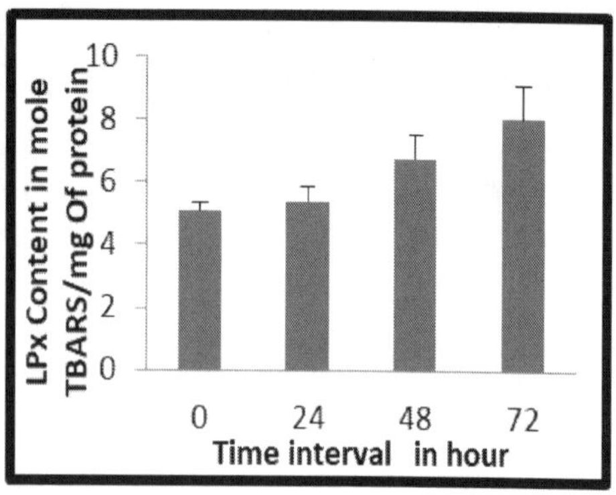

Figure 9.12.: Comparison of LPx in *Catla catla* Exposed to $Pb(NO_3)_2$ (2.658mg/L) different at Time Interval.

ii) Lead Nitrate

Lipid peroxidation (n mole of TBARS/mg Protein) in gillof *Catla catla* was 5.086±0.003 in control and after exposed with Lead Nitrate were 5.384±0.004, 6.746±0.453, 8.024±0.001 at 24 hour, 48 hour and 72 hour respectively (Figure 9.12). One way ANOVA was performed in order to analyse the effect of Lead Nitrate on LPx at different time interval in *Catla catla*.One way ANOVA revealed that the LPx level at different time interval is significant [$F(3,19)= 18.600$, $P=0.000$]. Post Hoc analysis revealed that Lipid peroxidation at different time interval when exposed withLead Nitrate in gill of *catla catla* were differ significantly significant at 0 hour, 24 hour,48 hour and 72 hour. ($P<0.05$; LSD).

Reduced Glutathione (GSH) Content

i) Cadmium Chloride

GSH content (mg/g tissue) in gill of *Catla catla* is 0.093 ±0.003 in unexposed and after exposed with Cadmium Chloride protein content were 0.098 ±0.008,0.101±0.004,0.100 ±0.004 at 24 hour,48 hour and 72 hour respectively (Figure 9.13). One way ANOVA revealed that the GSH content at different time interval is significantly different [$F(3,19)=10.418$, $P=0.032$]. Post Hoc analysis revealed that mean difference of GSH content at different time interval when exposed with cadmium Chloride in gill of *catla catla* was significant in control, 24 hour,48 hour and 72 hour($P<0.05$;LSD)

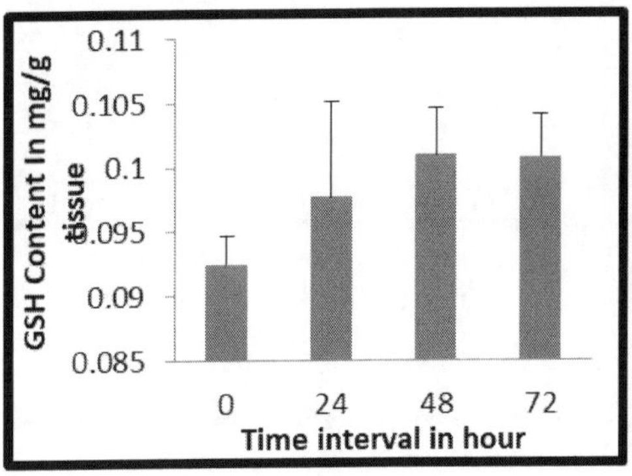

Figure 9.13: Comparison of GSH Level in *Catla catla* Exposed to $CdCl_2$ (0.453mg/L) at different time interval.

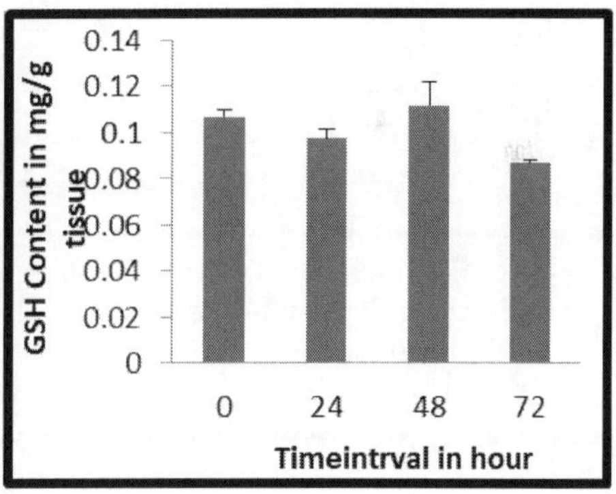

Figure 9.14: Comparison of GSH Level in *Catla catla* Exposed to $Pb(NO_3)$ at different time interval.

ii) Lead Nitrate

GSH content (mg/g tissue) in gill of *Catla catla* is 0.107±0.003 in control and after exposed with Lead Nitrate were 0.097±0.004, 0.111±.009,0.866±0.002 at 24 hour, 48 hour and 72 hour respectively (Figure 9.14). One-way ANOVA revealed that the GSH content at different time interval differ significantly [$F(3,19)=13.921$,

P=0.000]. Post Hoc analysis revealed that mean difference in GSH content at different time- interval when exposed with Lead Nitrate in gill of *catla catla* was significantly different in control, 24 hour,48 hour and 72 hour ($P<0.05$; LSD).

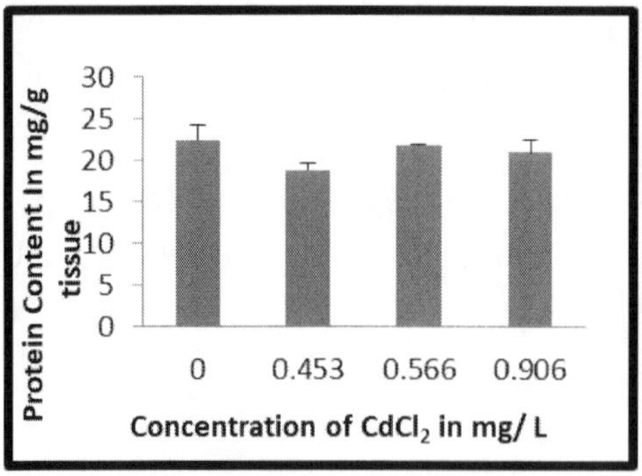

Figure 9.15: Comparison of Protein Content in *Catla Catla* exposed to Different Conc. of $CdCl_2$ at 24 Hour.

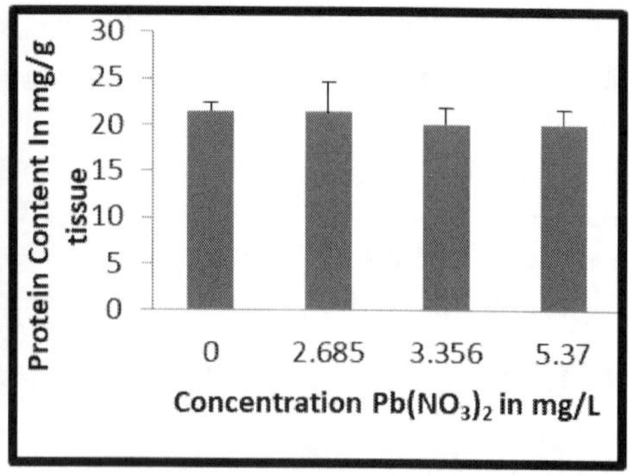

Figure 9.16: Comparison of Protein Content in *Catla catla* Exposed to different Conc. of $Pb(NO_3)_2$ at 24 Hour.

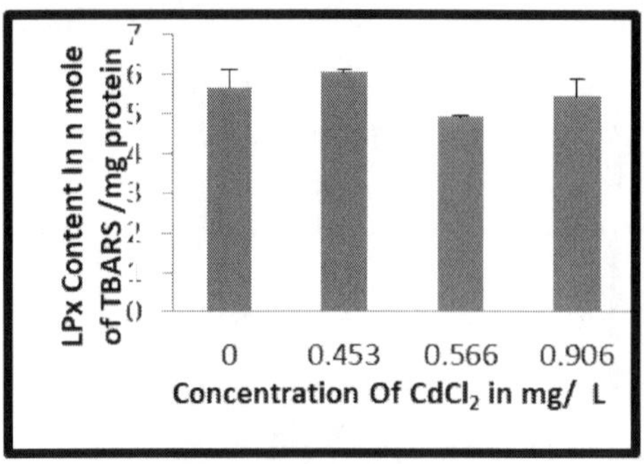

Figure 9.17: Comparison of LPx in *Catla catla* Exposed to different Conc. of $CdCl_2$ at 24 Hour.

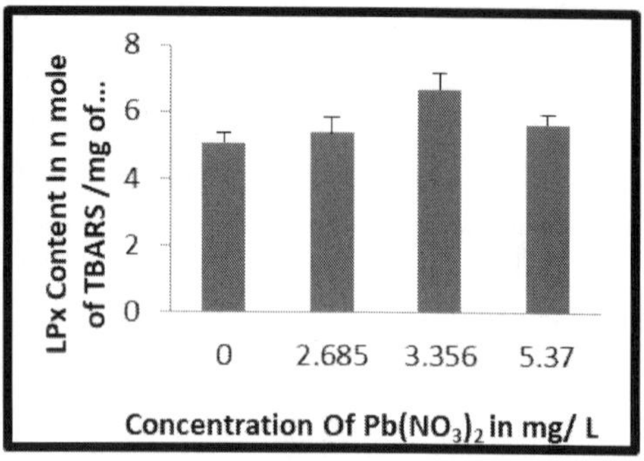

Figure 9.18: Comparison of LPx in *Catla catla* Exposed to different Conc. of $Pb(NO_3)_2$ at 24 Hour.

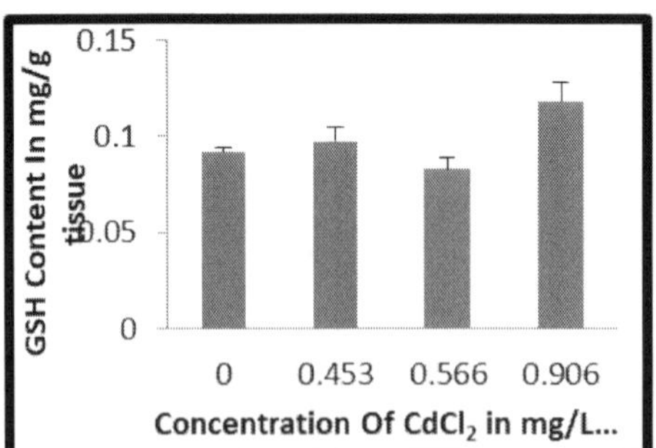

Figure 9.19: Comparison of GSH Level in *Catla catla* Exposed to different Conc. of $CdCl_2$ at 24 Hour.

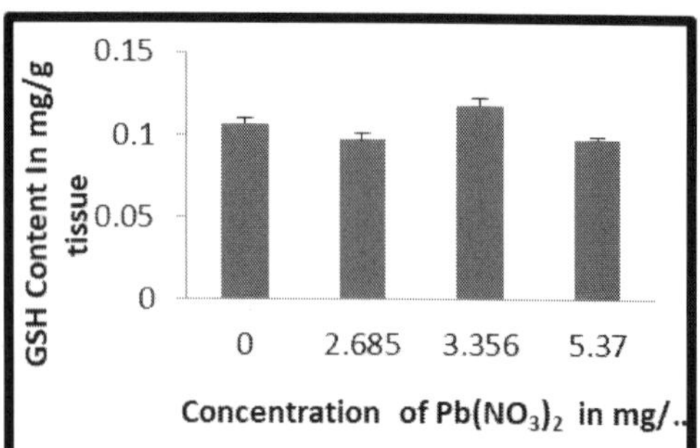

Figure 9.20: Comparison of GSH Level in *Catla catla* Exposed to different Conc. of $CdCl_2$ at 24 Hour.

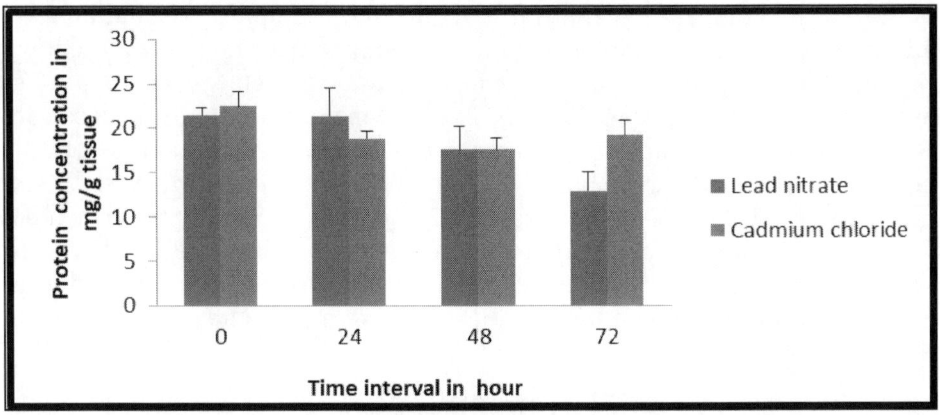

Figure 9.21: Comparative Study of Effect of Pb $(NO_3)_2$ and $CdCl_2$ on Protein Content of *Catla catla* at different Time Interval.

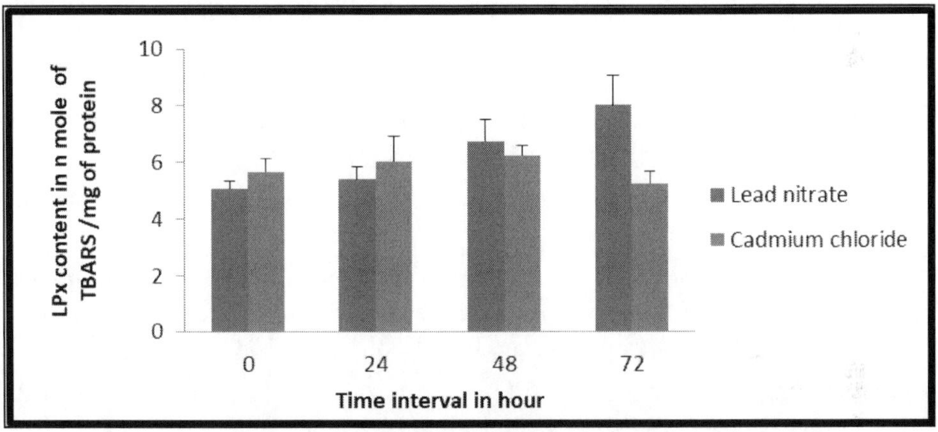

Figure 9.22 : Comparative Study of Effect of $Pb(NO_3)_2$ and $CdCl_2$ on LPx of *Catla catla* at different Time Interval.

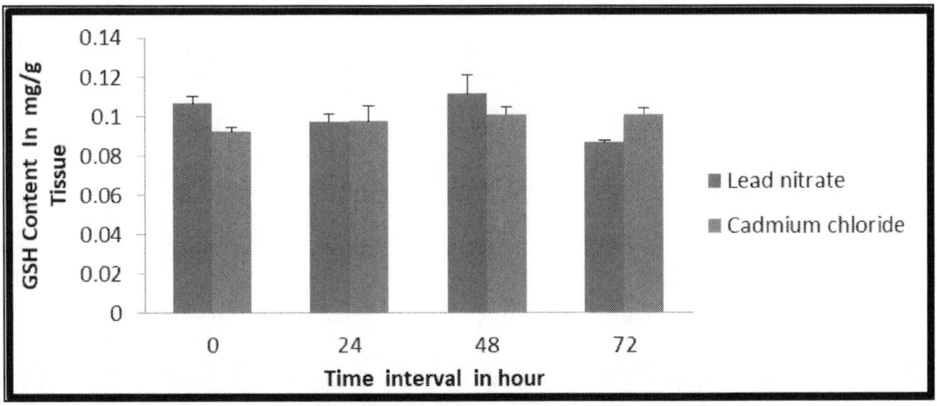

Figure 9.23: Comparative Study of Effect of $Pb(NO_3)_2$ and $CdCl_2$ on GSH Content of *Catla catla* at different Time Interval.

From the comparative study, it is found that the lipid peroxidation level increases at different time interval both in response to lead nitrate and cadmium chloride in comparison to that of control; the level was slightly decreased at 72h in case of cadmium (Figure 9.22). The GSH level increases at 48 h and decreases at 72h (Figure 9.23) and the protein content decreases with increase of time intervals (Figure 9.21) in response to lead nitrate when compared with that of control. Similarly, GSH level increased at 72 h (Figure 9.23) and protein content increased at 72 h (Figure 9.21) in response to cadmium chloride. There was an overall decrease in protein content may be due to impairment of protein synthesis. Increase of GSH content may help to adjust the toxic effect of lipid peroxidation in response to heavy metals.

REFERENCES

Beijere K and Jernelov A (1986) Sources, transport and transformation of metals in the environment. Handbook on the toxicology of metals. Elsevier, Amsterdam. pp 68-84.

Ellmanm GD (1959) Tissue Sulphhydryl groups. Arch Biochem Biophysics.

Lowry OH, Rosenbrough NM, Farr AL and Randall RJ (1951) Protein measurement with Folin phenol reagent. J Biol Chem 193: 265–275.

Ohkawa H, Ohisi N and Yagi K (1979) Assay of lipid peroxides in animal tissues by thiobarbituric acid reaction. Anal Biochem 95: 351-358.

Radhaiah V, Girjia M and Rao KJ (1987) changes in selected biochemical parameters in the kidney and blood of the fish, *Tilapia mossambica* (Peters), exposed to heptachlor. Bull. Environ. Contam. Toxicol. 39: 1006-1011.

Romeo M, Siau Y, Sidoumou Z and Gnassia-Barelliln M (1999) Heavy metal in different fish species from the Mauritania coast. Sci. Total Environ. 232: 169-175.

Wills ED (1969) Lipid peroxide formation in microsomes, General consideration. Biochem. J 113: 315-324.

Chapter 10

Traditional Agroforestry Systems: Maintaining Biodiversity and Conserving Natural Resources in Coastal Ecology of Odisha

U.S. Nayak[1], G. Shial[2], C.K. Das[1] and M.C. Behera[3]

[1]*Regional Research and Technology Transfer Station, Ranital, OUAT–756111, Odisha*
[2]*Krishi Vigyan Kendra, Bhadrak, Ranital, OUAT–756111, Odisha*
[3]*College of Forestry, OUAT, Bhubaneswar–751003, Odisha*
e-mail: gayadharbalasore@gmail.com

Introduction

Agroforestry system of land use is rapidly gaining importance in the recent years for meeting the dual objectives of achieving food security and environmental sustainability. It not only improves farm production and income but also conserves the natural resources, and local biodiversity and ensures their optimum utilization. This type of land use has been instrumental in reducing the CO_2 emissions from agro-ecosystems, stabilizes the atmospheric CO_2 concentration and hence minimizes the impact of climate change. The role of forests (or trees) in carbon cycles is well recognized (Singh and Lal, 2000) and they are a large sink of carbon

(Wang *et al.,* 2001). Rapid deforestation in the rural landscape through shifting cultivation, expansion of agriculture, infrastructure development like roads, railways, ports, mining, and industrialization left agroforestry is the only viable option left for mitigating the impact of climate change while providing food and livelihoods to the millions of smallholder coastal communities. The planting of trees along with crops improves soil fertility, controls and prevents soil erosion, controls water logging, checks acidification and eutrophication of streams and rivers increases local biodiversity, decreases pressure on natural forests for fuel, and provides fodder for livestock (Makundi and Sathaye, 2004).

Agroforestry is the judicious integration of woody perennials (trees, shrubs, palms, bamboo *etc.*.) with crops and/or animals on the same unit of land with some form of spatial arrangement or temporal sequence for meeting the multiple needs of the rural community like food, fodder, firewood, timber, medicine, and other ancillary benefits. Agroforestry is a shining approach that merges century's knowledge with modern science in a system and the concept of thinking small scale to achieve potentially big and transformative outcomes (Steiner, 2012). A wider definition of agroforestry encompasses a variety of practices, including trees on farm boundaries, trees grown in close association with village rain water collecting ponds, crop fallow rotations and a variety of agro-forests, silvipastoral system and trees within settlements (Pandey, 2007). It facilitates the nutrient recycling, improves the soil fertility, arrests soil erosion and land degradation, conserves soil and water, improves village environment, conserves biodiversity and serves as a major carbon sink. This system is ecologically more complex and both genetically and structurally diverse than the mono-cropping system and hence, more stable and resilient to biotic and abiotic stresses. Realizing the importance of this type of land use, emphasis is being given to promote agroforestry in agricultural lands, permanent fallow lands, wastelands, community pastures and other common property resources for stabilizing food production, sustaining the livelihoods and extending the environmental services. Besides, agroforestry is the best alternative land management system to meet the escalating demand for fodder, firewood and timber of the farming communities under the scenario of rapid shrinkage of pasture and community wastelands in the rural areas. Higher income and productivity, increase in fuel wood, population pressure, declining land availability and soil conditions have led farmers to adopt agroforestry (Panda *et al.,* 2010 and Tekale *et al.,* 2020). Agroforestry systems have potential to produce100 million cubic meter timber/pulpwood for industrial and domestic use (Chavan *et al.,* 2015), which fulfills the 65percent of the country's timber demand, two third of small timber demand, 70–80 percent of plywood demand and 60 percent raw material demand for paper pulp. The agroforestry systems in India include trees on farms, community forestry and a variety of local forest management and ethno forestry practices (Pandey, 2007). In India, the practice of growing scattered trees on farmlands is quite old and has not changed much over centuries; these trees are multipurpose, used for shade, fodder, fuel wood, fruit, vegetables and medicinal uses (Murthy *et al.,* 2013).

Traditional Agroforestry Systems

However, even before the evolution of agroforestry as a concept and technology, farmers are maintaining a variety of tree species in their backyard home garden and boundary of crop fields along with their agricultural crops since time immemorial. It has been evolved along with the development of human civilization and continues to play a vital role in maintaining the sustainability of rural livelihoods. This is considered as the oldest form of tree crop combination and sometimes referred as ethnoforestry, homestead forestry, backyard forestry and forestry outside the natural forest and consists of a diverse plant species to meet the basic needs of the rural households. It has been a common practice of growing trees, shrubs, vines and herbs in or around the home stead to make the rural people self-sufficient in the basic needs like food, firewood, fodder and shelter. The farmers have also the practice of growing and maintaining trees like *Acacia nilotica, Acacia auriculiformisis* and *Borassus flabellifer* on the farm bunds. These plants are pruned regularly so that their shade does not affect the yield of annual and the pruned branches are used as staking materials for growing vegetable crops or as green manures and fire wood. These traditional agroforestry systems are deeply linked with the lifestyle and livelihoods of the resource poor small holders, insulate them from the climatic uncertainties and have immense social and cultural values. However, these are not just a group of trees of different species, but is a valuable ecosystem in itself and offers a wide range of eco-system services including the supporting services, provisioning services, regulating services, landscape beauty, aesthetic values and spiritual wellbeing. It also provides a valuable niche for the diverse flora, fauna and micro-organisms including the multipurpose tress, valuable herbs, pollinators, predators, decomposers, nutrient solublisers. Besides, it provides protection to farm and household from wind (shelter belts), shelter to birds and other beneficial fauna and act as live fences to prevent the entry of stray animals. The perennial vegetation in the agroforestry system provide alternate habitat and food supplements to the beneficial arthropods like pollinators, predators and parasitoids and therefore, strengthen the valuable regulating ecosystem services which is critically essential for the sustainability of home garden and agriculture landscape. It has been often considered as an important site for domestication of the wild and underutilized trees and crops through trial and error method, species selection and indigenous management practices and is a valuable place for in-situ conservation of precious genetic resources. Rural community mostly meets their firewood and timber demand from their home gardens and therefore, minimizes the pressure on the forest land. They also even provide opportunities for supplementing the family income through sales of poles, timber and fruits in the normal years and more importantly during the years of crop failure and other natural calamities. In spite of the very small average size of the management units, home gardens have a high species diversity and usually 3-4 vertical canopy strata of woody and herbaceous species (Nair, 2008).This system of land use is a self-sustaining production system that largely avoids the use of external inputs and based on the principle of resource recycling and optimum utilization of family labour. The promotion of sustainable agroforestry practices on a large scale in future only possible through amalgamation of proactive farmer policies of government, involvement of the industries, support services from NGOs and willingness of farmers (Verma *et al.,* 2017).

Traditional Agroforestry Systems in the Coastal Ecology of Odisha

These agroforestry systems are specific to agro-climatic zone, agro-ecological situation and social set up of the community. The diversity, and species composition of the systems largely depend on the topography, soil and climatic situation, family requirements, traditional beliefs, local preferences and the indigenous knowledge about use of the species. Therefore, variety of tree species are grown and maintained traditionally by the community of different regions of the country for deriving their multiple needs and supporting the local biodiversity. In Odisha diverse traditional agroforestry systems exist in different areas characterized with high species diversity including multipurpose tree species, shrubs and palms. While, in the drought prone western parts *Acacia nilotica* and *Butea monosperma, Cassia siamea, Azadirachta indica,* based farm forestry are the dominant systems, in the central inland districts *Madhuca latifolia* and *Scleichera oleosa* based agroforestry are more prevalent. However, in the coastal areas tree species like Palms, *Albizia lebbek, Gmelina arborea, Pongamia pinnata, Erythrina indica, Samanea saman, Terminali* sp, *Bambusa stricta and Phoenix acaulis etc.* are maintained by the farmers along with their agricultural crops in their backyard as home garden or in the field as farm forestry. Agroforestry in the coastal areas has the potential to improve the livelihoods of vulnerable rural communities by supplementing their household income, while increasing their resilience to climate risks. Since, the ancient time the multipurpose perennial legumes like *Albizia lebbek, Gmelina arborea, Erythrina indica, Acacia nilotica, Dalbergia sissoo, Bauhunia purpurea etc.*. have been managed in the home garden for timber and fire wood purpose, to maintain soil fertility and feed livestock during the lean season. Traditional fruit species like Bael (*Aegel marmalous*), Jack fruit (*Artocarpus integrifolius*), Jamun (*Syzygium cumuni*), Custard apple (*Annona squamosa*), Elephant apple (*Dillenia indica*), Carambola (*Averrhoa carambola*), Wood apple (*Feronia elephantum*), Ber (*Ziziphus mauritiana*), Star gooseberry (*Cicca disticha*), Biggord orange (*Citrus aurantinum*) *etc.*. were maintained in the backyard to take care the food and nutritional security of rural households. In addition to their rich nutritional values, these minor fruit crops have been playing significant role in Indian system of traditional medicine and widely used against many common diseases. Neem (*Azadirchta indica*), Pongam (*Pongamia pinnata*) have been the integral part of traditional agroforestry system for their medicinal and insecticidal values. Different species of bamboos are still essentially important for the construction and maintenance of rural houses, as a source of nutritious fodder and food during the lean season. The traditionally palm based (coconut, *Cocos nucifera* and palm, *Borassus flabellifer*) agroforestry are still preserved in the coastal area for food during lean season, fire wood and building materials. Generally, the backyard home gardens are multi-tier systems characterized by a high density and diversity of plant species consisting of trees/ palms, shrubs, climbers and seasonal crops that ensures efficient utilization of nutrients, water, sunlight and space. Common annual crops like papaya, banana, yam, arrowroot, turmeric and ginger are generally grown under partial shade of the tree canopy of coconut based agroforestry system in coastal areas (Isaac and Nair, 2000 and Maheswarappa *et al.*, 2000). Integration of livestock (cow, buffalo) and poultry birds (hen, ducks) are also observed in the coconut based agroforestry system for increasing system productivity and better resource recycling (Sahoo, 2007 and Ahmed and Hazarika, 2007). Coconut based agroforestry system are playing multiple roles like various productive, protection and ameliorative, recreational and educational as well as developmental role (Panda *et al.*, 2020)

Biodiversity in the Traditional Agroforestry Systems

Realising the priceless importance of traditional agroforestry system an attempt is made to analyze its species diversity, bio-physical aspects, resource benefits and management practices in the coastal eco-systems of Odisha state of India. The study was conducted in the Bhadrak and Balasore districts coming under the North Eastern Coastal Plain Agro-climatic zone of Odisha. The common indigenous agroforestry systems practiced by the local people were studied with reference to species diversity and multipurpose use. It was revealed from the study that the size of traditional agroforestry systems varies from 0.04 to 1.20 ha with an average size of 0.2 ha and among the species fruits are the dominant tree component follows by forest plants, bamboos, palms. During the investigation a distinct variation in size of the system, species diversity, distribution and composition were observed. There are four major traditional agroforestry systems *i.e.* Coconut based, Palm based, Bamboo based and babul based (Table 10.1), found in the coastal areas and among them the Coconut based system is the dominant one with the maximum system frequency. In coconut-based agroforestry system coconut occupies the top storey, mango jackfruit, gambhar are maintained in the third storey, drum sticks, papaya, lemon, custard apple are managed in the second storey and annuals like ginger, turmeric, pineapple, cow pea, brinjal and other vegetables are grown in the first storey in the agroforestry. The palm-based system is mostly practiced in the paddy lands adjacent to the homestead. The farmers maintained the palms in the field bunds, while growing rice, pulses, oilseeds and vegetables in the crop land and the homestead is having diversity of annual and perennial plants. Bamboo based system is also quite popular in some parts of the coastal areas harbouring variety of trees and crops. The Babul based system is dominant in the coastal saline and low-lying tracts and significantly contribute towards better resilience and economic wellbeing the rural community. The plant species maintained in the traditional agroforestry systems have multipurpose use as food, fodder, firewood, timber, medicine and other specific uses as highlighted in Table 10.2.

Table 10.1: Common Traditional Agroforestry Systems of the Coastal Odisha

Traditional Agroforestry Systems	*Dominant Species*	*Secondary Species*	*Frequency of the System*	*Advantage*	*Constraints*
Coconut (*Cocus nucifer*) based	Coconut, Mango, Albizia, Guava, Jackfruit	Gambhar, Bauhunia, Dillenia, Carmbola Moringa, Papaya, Lemon, Pine apple, vegetables, Ginger, turmeric.	43 per cent	✰ Adaptable to homestead land of coastal region. ✰ Multiple use in food, house hold materials like brooms and baskets, timber building material, fodder and green manure	✰ Low productivity due to improper space management and lack of improved production technology. ✰ Susceptible to high wind situation and attack of Eriophid mite.
Palm (*Borassus flabellifer*) based	Palm, Rice, pulses Neem, Arjuna	Albizia, Custard apple, Papaya, Bael, Date palm and vegetables	22 per cent	✰ Grown on the field bunds and tolerant to waterlogged condition ✰ Multiple use in off-season food, building material, household items like baskets, fencing pools, timber, fodder and green manure	✰ Vulnerable to high wind and lightening condition ✰ Lack of technology for better production of jaggery
Bamboo (Bamboo sps) based	Bamboo, Guava, Mango	Pongam, Samania, Tamarind, Jackfruit, Erythrina, Elephant apple, Lemon, Papaya, Moringa and vegetables	18 per cent	✰ Adaptable to marginal lands and tolerant to saline and waterlogged condition ✰ Protect the land from soil erosion	✰ Low productivity due to improper space management and lack of scientific propagation technique

Traditional Agroforestry Systems	*Dominant Species*	*Secondary Species*	*Frequency of the System*	*Advantage*	*Constraints*
				☆ Multipurpose use in building material, off-season fodder, household and agricultural purposes like making basket and hand fan, sieving and winnowing	☆ Gradual low demand of bamboo products
Babul (*Acacia nilotica*) based	Acacia, Rice, Pulses, Arjuna	Neem, Mango, Guava, Sponias, Albizia, Black berry, Moringa, Papaya and vegetables	12 per cent	☆ Specific adaptability to coastal saline area ☆ Multiple use as fodder during lean season, agriculture and small household implements, building material and fencing pools, soil fertility enrichment	☆ Over exploitation for fencing and fodder ☆ Poor preference due to thorny branches

Table 10.2: Common Plant Species in the Traditional Agroforestry Systems of Coastal Areas and their Common Use

Common Use	*Common Plant Species*
Food	Coconut, Palm, Mango, Guava, Jack fruit, Wood apple, Elephant apple, Carambola Spondias, Black berry, Tamarind, Dillenia, Custard apple, Country gooseberry, date palms, Lemon, Lime, Papaya, Moringa, Pine apple
Timber	Albizia, Babul, Samania, Jack fruit, Mango, Black berry, Arjuna, Neem
Fire wood	Babul, Albizia, Samania, Tamarind, Arjuna, Manial Tamorind
Fodder	Manial Tamorind, Bauhinia, Albizia, Babul, Bamboo
Green manure	Bauhinia, Babul,
Live fence	Erythrina, Bombax, Bamboo, Jatropha, Adhatoda, Lawsonia, Pandanus, Euphorbia, Vitex, Thevetia
Medicinal	Adhatoda, Bael, Neem, Lawsonia, Tulsi, Nictanthus, Vitex, Black berry, Country gooseberry
Tree borne oilseeds	Neem, Pongam, Polang

Conclusion

Traditional agroforestry system has strong interwoven connection with food security, livestock rearing, livelihood pattern, socio-economic set up, religious and spiritual values and environmental benefits of rural societies and therefore, has been considered as a unique resource management system and model agro-ecosystem with typical characteristics of diversity, stability and resource use efficiency. However, over the years these traditional agroforestry systems are rapidly disappearing in all the eco-systems in general and coastal environment in particular due to high population pressure and subsequent land fragmentation and diversion of land for habitat purpose. This high diversity-based land use system has been replaced by monocropping of few selected tress or annual crops undermining the sustainability of rural food, nutrition and livelihood security. Many of the plant species are rapidly losing their significance and are in the verge of extinct making the coastal ecology more vulnerable to climate extremes. Therefore, it has attracted the attention of researchers, development practitioners and policy makers during past few years for their systematic conservation and productivity enhancement through holistic approach. However, there has been a growing realization of the benefits of tree crop integration systems in the coastal areas. Conservation and management of traditional agroforestry systems are of immense significance in stabilizing the food production, economic benefits, social set up and rural environment. It helps to improve the productivity marginal lands while conserving the agro-biodiversity and sustaining the rural economy.

REFERENCES

Ahmed AA and Hazarika DN (2007) Agroforestry systems and practices prevailing in Assam. Agroforestry: systems and practices (eds. Puri, S. and Panwar, P.), pp. 347-355.

Chavan SB, Keerthika A, Dhyani SK, Handa AK, Newaj R and Rajarajan K (2015) National Agroforestry Policy in India: a low hanging fruit. Current Science 108(10): 1826-1834.

Isaac SR and Nair MA (2000) Coconut gardens for agroforestry. Indian Coconut Journal 31(3): 14-16.

Maheswarappa HP, Hegde MR and Nanjappa HV (2000) Arrowroot (*Maranta arundinacea* L.): a potential intercrop in coconut garden. Indian Coconut Journal 31(3): 20-21.

Makundi WR and Sathaye JA (2004) GHG mitigation potential and cost in tropical forestry-relative role for agroforestry. Environment, Development and Sustainability 6: 235-260.

Murthy IK, Gupta M, Tomar S, Munsi M and Tiwari R (2013) Carbon Sequestration Potential of Agroforestry Systems in India. Journal of Earth Science and Climate Change 4: 131.

Nair PKR (2008) Structure and function of Home gardens. An Introduction to Agroforestry (eds. Nair PKR), pp. 91-94.

Pandey DN (2007) Multifunctional Agroforestry Systems in India, Current Science 92 (4): 455-463

Panda NK, Sarangi SK, Das HK and Kar MR (2020) Role of coconut (*Cocos nucifera*) based agroforestry system in coastal Odisha. Journal of Pharmacognosy and Phytochemistry 9(4): 1742-1745.

Sahoo UK (2007) Agroforestry systems and practices prevailing in Mizoram. Agroforestry: systems and practices (eds). Puri S and Panwar P, pp. 367-383.

Singh R and Lal M (2000) Sustainable forestry in India for carbon mitigation. Current Science 78: 563-567.

Steiner A (2012) Agroforestry and Transition to the Future. In: Agroforestry-The future of Global land use (eds). Nair PKR and Garrity D, Springer, Dordrecht, pp17-27.

Tekale VS, Yogesh NJ and Thakare PN (2020) Knowledge and adoption of paddy based agroforestry practices. Journal of Pharmacognosy and Phytochemistry 9 (2S): 69-72.

Verma P, Bijalwan A, Dobriyal MJR, Swamy SL, Thakur TK (2017) A paradigm shift in agroforestry practices in Uttar Pradesh, Current Science 112(3): 509-516.

Wang X, Feng Z and Ouyang Z (2001) The impact of human disturbance on vegetative carbon storage in forest ecosystems in China. Forest Ecology and Management 148: 117-123.

Chapter 11

Studies on Commercial Host Plants and Prevalence of Natural Enemies (Predators and Parasitoids) of Lac Insect *Kerria lacca* (Kerr.) around the Kuldiha Wildlife Sanctuary, Odisha

G. Shial[1], U.S. Nayak[2] and K. Upadhyay[3]

[1]*Krishi Vigyan Kendra, Bhadrak, OUAT, Odisha*
[2]*Regional Research and Technology Transfer Station, Ranital, OUAT, Odisha*
[3]*School of Earth Science and NRM, Mizoram University*
e-mail: gayadharbalasore@gmail.com

Introduction

Forest is an important sector offering a magnificent enrichment in the livelihoods of tribal people through gainful occupation and NTFPs based self-employment in the Indian economy. Since time immemorial forest fringes and people in rainfed areas depends on these forests for their sustenance to a great extent. These non-wood forest produce (NTFPs) provide alternate source of food and income to sustain millions of tribes (Ajaz-ul-Islam *et al.*, 2013). India earns more than 50 per cent of the forest revenues and 70 per cent of the foreign exchange from NTFPs (Shiva, 1993). Collection and sales of NTFPs play a significant role in poverty alleviation than the subsistence, seasonal and insecure agricultural sector.

There is a greater role of NTFPs in socio cultural and conventional culture of forest inhabitants, particularly the tribal, landless, women and other underprivileged rural people as perceived by Prasad (2011). Among these resins and gums play an important group of products in the baskets of NTFPs that explore high price at national and international markets (Mulugeta *et al.,* 2003). It is also generally specified as a Non-Timber Forest Produces (NTFPs) (Jaiswal *et al.,* 2006). Lac is most preferable NTFP perceiving its lucrative price cultivated on various host trees that has potential to improve the livelihoods of households of small and marginal farmers and forest dwellers. It is one of the most valuable gifts of nature to mankind, the only resin of animal origin actually the secretion by a tiny scale insect, belonging to the family Tachardiidae, superfamily Coccoidea of the order Hemiptera (Pal, 2009; Mohanta *et al.,* 2012 and 2014). Lac is secreted as a protective covering throughout the body of lac insects except mouth, two brachial pores and anus. The English word "lac" in Hindi is derivative of Sanskrit word, 'Laksha', meaning a lakh or hundred thousand due to innumerable number of insects in small patches indicating the gregarious habit of this insect (Krishnaswami, 1962). Vedic people of ancient India knew that the lac is obtained from numerous insects (Kerr, 1782).

Jaiswal *et al.* (2006) stated that share of lac cultivation is around 28 per cent of total agriculture income. Lac is one of the most valuable gifts of nature to man and providing regular income in absence of other cash crops particularly in rain fed areas and conserve the associated fauna and flora for posterity. In several states lac cultivation is an efficient activity for enhancing livelihood of sub-forest, forest and rainfed areas (Jaiswal *et al.,* 2012). Lac based farming is a low input natural resource-based enterprise provides additional income and developed the socio-economic conditions of lac farmers (Jaiswal *et al.,* 2006, Pal *et al.,* 2010). Shellac is a processed product of lac is extensively used in electronics, chemicals, defence, food and other industries (Chen, 2005). Sharma (2013) stated that lac has significant use in paints and polishes, printing, fruit coating, pharmaceuticals, cosmetics, and many other similar applications. Moreover, the large-scale demand for natural products in globally enhances wider application of lac products in food, cosmetics, pharmaceuticals, varnishes and paints (Sharma and Ramani, 2011). Due to its non-injurious, durability and wonderful adhesive characteristics it has versatile uses in industrial fields of food, pharmaceuticals, paints, adhesives, jewellery, cosmetics, perfumes, polishes, varnishes and in textile dye (Srivastava and Jeet, 2020).

Lac by-products are biodegradable, non-toxic, eco-friendly and have multipurpose uses and excellent export ability (Chattopadyay, 2011). Sharma *et al.* (2006) also proposed that adoption and inspiration of lac cultivation eliminate environmental degradation, maintain the ecological balance and also preserve endangered lac insects, associated fauna and flora (Sharma *et al.,* 2006). According to him this lac insect-wide host range association act as a tool for the protection of biodiversity *viz.,* various soil flora, fauna and soil microorganisms. Lac insect is one of the economically most important insects and produces lac, a valuable minor forest produce (Shah *et al.,* 2015), cash crop (Ramani *et al.,* 2007) and an export-oriented product (Anon, 2005). Lac production and productivity depends upon availability of host plants, their management and quality of brood use, poor institutional linkages, climate change and extremist activities (Shah *et al.,* 2015, Bharat, R, 2010), live lac insects settled (Khobragade *et al.,* 2012, Rathore 2011, Jhangel *et al.,* 2013, Sharma *et al.,* 2015, Ghugal *et al.,* 2015, Namdev *et al.,* 2015, and Sahu *et al.,* 2016) till the harvesting of the lac crop. Ghosal, 2013 also described that lac insects thrives on tender branches of lac host plants. Lac cultivation helps to build up the plant–insect interactions (Shah, 2016) and at present around 5 per cent of lac hosts are used for lac cultivation (IINRG 2018). In India, conventional major host plants are Butea monosperma (Palas), *Schleichera oleosa* (Kusum) and *Ziziphus mauritania* (Ber) (Ghosh *et al.,* 2014; 2016). *B. monosperma* contributes highest source of secondary income for tribals accounting for 46.9 per cent followed by *S. oleosa* 32.3 per cent and *Z. mauritiana* 20.8 per cent (Jaiswal *et al.,* 2006). Due to lack of adequate knowledge on lac cultivation only 5 per cent of host plants of *S. oleosa* and *Z. mauritiana* are utilised for producing Kusmi strain of lac. While only 1 per cent of plants of *B. monosperma* for lac cultivation (Das and Kumar, 2013). *B. monosperma, S. oleosa, Ziziphus mauritiana, Ficus* spp., *Cajanus cajan* and *Flemingia semialata* are the major host plants of *Kerria lacca* (Sharma and Ramani, 2010; Monobrullah *et al.,* 2016).

Sarvade *et al.* (2018) reported that throughout the world 09 genera and 99 species of lac insects, out of which 2 genera and 26 species are found in India. *K. lacca* (Kerr.) is the most widely and commercially important Indian lac insect found in tropical or subtropical regions. Varshney (1976) and Ramani *et al.* (2007) also reported that the species of *Kerria* are found across the India but in scattered locations. Lac insects are phloem sap suckers and flourish on more than 400 host plants around the world (Ahmad *et al.,* 2012; Ramani *et al.,* 2008). Some researchers have stated that these sap sucking insects thrives well on a range of 250 host plants (Sharma *et al.,* 2006: Singh *et al.,* 2009; Pal, 2009; Mohanta *et al.,* 2012; Shah *et al.,* 2018).

Lac crop is susceptible to various types of insect predators, parasitoids and diseases like other agricultural crops due to sedentary nature of insects where they permanently attached to the host plants for sucking nutrition (Varshney, 1976). These Insect pests cause economic loss to the infested lac crops due to decline in lac production. Mohanasundaram *et al.* (2016) reported that parasitoids alone constitute 93 and 89 per cent population in baisakhi crop during 2011-12 and 2012-13 among lac associated fauna, followed by predators and hyper parasitoid on ber. Chowdhury *et al.* (1971) described that the density of population of lac insects determined the extent of parasitisation by the Chalcid parasites correlated with lac. Consumption of host plants by herbivores also influenced the growth and development of parasitoids (Eben *et al.,* 2000). Ogle *et al.* (2006) also suggested that the wide range of predators and parasitoids associated in lac production. Genetic resources of the lac insect available in India are under threat as many lac insects and associated fauna have been abandoned or its habitat destroyed due to the disappearance of a substantial number of local populations (Mohanta *et al.*). Chowdhury *et al.* (1971) described that the density of population of lac insects determined the extent of parasitisation by the Chalcid parasites correlated with lac. Consumption of host plants by herbivores also influenced the growth and development of parasitoids (Eben *et al.*, 2000). Ogle *et al.* (2006) also suggested that the wide range of predators and parasitoids associated in lac production. Genetic resources of the lac insect available in India are under threat as many lac insects and associated fauna have been abandoned or its habitat destroyed due to the disappearance of a substantial number of local populations (Mohanta *et al.*, 2014).

Among the lac growing states, Jharkhand state is the 1st in production followed by Chhattisgarh, Madhya Pradesh, Maharashtra and Odisha. These five states contribute around 95 per cent of the national lac production. Share of Jharkhand in national lac production is about 58 per cent followed by Chhattisgarh (16.1 per cent), Madhya Pradesh (11.9 per cent), Maharashtra (5.6 per cent) and Odisha (3.2 per cent) (Yogi *et al.*, 2015; 2018). Odisha formerly called Orissa is the 8th largest state by area and 11th largest by population of the country. Interestingly, this state is blessed with ample of forest biodiversity and Non-Timber Forest Produces (NTFPs). Forest cover of the state is 51,619 sq km which constitutes 33.15 per cent of the state's geographic area (Forest Survey of India, Report 2019). Lac culture is traditionally practiced in some pockets of Balasore district of Odisha, indicating that agro-climatic condition of region is suited for lac cultivation. Nilgiri block of Balasore is historically a major lac producing area in the North Eastern India. Most of the villages this block has abundant host trees naturally growing on the on borders of paddy fields (bunds) and barren lands in clusters. All three major host trees such as Ber, Kusum and Palash are found available in these areas. Some of the households especially tribal people are quite having habitual access and interest to make use of host trees for lac cultivation and followed general lac cultivation process steps. As lac is cultivated widely in different parts of Balasore district and can be better exploited for commercial production. But with improvement in Agriculture, indiscriminate deforestation and vast industrialization, the naturally existing host trees are depleting more rapidly. Moreover, there is non-availability and scanty knowledge among lac cultivators about the conservation of these of their natural habitat, lac insect enemies interactions and causes for its occurrence are unknown. There is no documented literature available about the availability of naturally occurring commercial lac host plants and the natural enemies of lac insect. Therefore, an attempt has been made under present investigation to record the naturally occurring commercial host plants and natural enemies of lac insect prevailing around the Kuldiha wildlife sanctuary, Odisha for future promotion of lac cultivation in the region.

Study Sites

The present study has been carried out in Nilgiri block of Balasore district, Odisha, India by establishing ten numbers of randomly selected sites of Garadihi

Gram Panchayat of Nilgiri block of Balasore district is known as Tribal Sub-plan Areas (TSP). All the sites are located on the undulating plain of Kuldiha foothill, blessed with rich floral and faunal diversity. It is located in adjacent to Kuldiha Wildlife Sanctuay. Kuldiha Wildlife Sanctuary, a part of mega-habitat bounded by the trijunction of Balasore, Mayurbhanj and Keonjhar districts, is covered with densely forested hill range spreading over 272.75 km^2. It is located in the South-western part of Balasore District under Nilgiri Civil Sub-Division in the State of Odisha. It lies between 21° 20' to 21° 30' North latitude and 86° 26' to 86° 45' East longitude. The area has a subtropical climate with three distinct seasons *i.e.* summer, monsoon and winter. The mean annual rain fall of study site is about 1630 mm and enjoys a tropical climate. Because of its proximity to Bay of Bengal, climatically this tract is comparatively humid- hot. The mean maximum temperature is 38° C during April – May and minimum 8° C during November – January. The maximum temperature at times falls beyond 42° C. Forest type is mostly tropical deciduous forest type (Champion and Seth, 1968).

The forests of Kuldiha are broadly divided into i) Northern tropical semi evergreen forests, ii) Peninsular (Coastal) Sal Forests, iii) Moist mixed deciduous forests and iv) Mixed Dry deciduous forests. There are 12 (twelve) revenue villages with 54 hamlets inside the Sanctuary. Besides this there are 66 (sixty-six) revenue villages all around the protected area. Villagers' dependent on the forest resources of the sanctuary to a great extent. Agriculture is the main occupation of tribal people. Collection of Non-Timber Forest Produces (NTFPs) is the next important occupation. They collect firewood, timbers, medicinal plants, leaves *etc.*. are from the forests. Villagers' residing at close proximity allow cattle into the sanctuary, which is a common practice in the area. Lac cultivation is also a subsidiary source of income of forest fringes. Soil types varying from sandy to heavy red sandy loam found in the area. The rock of the site consists of Khondolites are represented by quartz, garnet, sillemanite, schist and garnetiferous quartzites. They are greyish to purple or reddish brown in colours. The constituent minerals are quartz feldspar, garnet, sillemanite and occasionally graphite. Basic pyroxene granites rocks are dark coloured and are generally fine to medium grained. Acid and Intermediate Charnockite rocks are fine to medium grained, greenish grey in colour. Garnet ferrous granite gneisses and granulites rocks are medium grained, grey in colour. A good exposure of laterite is found in this area in the south west portion. This is an alteration product of khondolites.

Methodology and Data Collection

The study is based on the primary data collected from five major lac growing villages namely Tiakata, Garadihi, Chaindar, Balichua and Chekamara of Garadihi Gram Panchayat to survey all the possible habitats such as agricultural field, homestead land, road side, village commons for recording the data for the year 2017-18 to fringe forests (*i.e.*, 5 Km) around Kuldiha reserve forest of Balasore, Odisha. To explore availability of naturally occurring commercial lac host plants and the natural enemies of lac insect survey was carried out. Through visual investigation with help of binoculars lac insect populations were noticed on reported lac host species. The branches and twigs having the lac insect encrustations were cut with help of secateurs and covered with moistened cotton plugs at both the ends to avoid desiccation. Samples were collected randomly from 10 plants each containing with 30 cm length after removal from the host plants. This was kept in 60 mesh nylon basket (bag) with for recording data on emergence of predators and parasitoids, if any. The method used by Jaiswal *et al.,* 1998 adopted for the study natural enemies of lac insects. To record the prevalence of natural enemies of lac insect lac infested branches and twigs brought to the lac insect museum cum laboratory of Dept. of Entomology, College of Agriculture, OUAT, Bhubaneswar and placed individually into glass vials plugged with cotton for about a month and the emerged predators and parasitoids. Lac associated fauna (parasitoids and predators) were collected from glass vials every day continuously up to one month and were got identified by them. Based on morphological characters of the predators, parasites and hyper parasitoids with the help of lac insect practical manual of Mohansundaram et. al., 2016 fauna that became visible were collected and separated for their identification under the microscope. Prior to undertaking the study, discussion was made with local authorities of forest and environment, Odisha. Thereafter, various sites were identified about lac insect and their host plant potential areas. Information was also gathered from lac traders and farmers in the blocks. With this information study sites were identified in the district.

Assessment of Major Conventional Lac Host Plants around Kuldiha Wildlife Sanctuary

Three types of major conventional lac host plants around Kuldiha wildlife sanctuary were studied (Table 11.1). Diversified plant communities are characterized by high properties of semi-natural habitats such as such as agricultural field, homestead land, road side, village commons *etc.*.

Table 11.1: Major Conventional Lac Host Plants around Kuldiha Wildlife Sanctuary

Sl.No.	*Common Name*	*Scientific Name*	*Family*	*Types of strain*	*Distribution*
1	Kusum	*Schleichera oleosa*	Sapindaceae	Kusumi	Forest, farmer's field and wastelands
2	Palash	*Butea monosperma*	Leguminosae	Rangeeni	Forest, field bunds, road side, waste lands and in uncultivated lands
3	Ber	*Zizyphus mauritiana*	Rhamnaceae	Kusumi and Rangeeni	Forest, field bunds, road side, waste lands and in homestead lands.

Strains of Lac Crops

The Indian lac insect, *Kerria lacca* (Kerr) has two different strains, Rangeeni and Kusumi and both the strain produces two crops in a year (bivolatine). Kusumi strain is superior (Kumar *et al.,* 2002; Sharma *et al.,* 2006) and it produces two crops, Jethwi (Jan-Feb. to June-July) and Aghoni (June-July to Jan-Feb.). Strains of lac insects are distinguished on the basis of differences in lifecycle, host plant preference and the quality of lac produced. *Rangeeni* is characterized by unequal duration of bivoltine life cycle (4 and 8 months) and whereas *kusmi* has equi-durational life cycle (6 months each). The crop-wise, different developmental stages of rangeeni and kusmi lac insect and duration have been given in Tables 11.2 and 11.3, respectively.

Table 11.2: Stages of Lac Insects and their Life Span in Days*

Insect (Strains)	*Crop*	*Stage*			*Male*		*Female*	
		I	*II*	*III*	*Adult*	*Total*	*Adult*	*Total*
Rangeeni	Rainy season (Katki)	20 (3)	14 (2)	8 (1)	2	44 (6-7)	67 (10)	109 (16)
	Summer (Baisakhi)	50 (7)	40 (6)	15 (2)	3	108 (15-16)	145 (21)	(250) (36)
Kusumi	Winter (Aghani)	18 (3)	13 (2)	10 (2)	2	43 (6-7)	150 (21)	193 (28)
	Summer (Jethwi)	30 (4)	24 (3)	10 (2)	2	66 (9-10)	90 (13)	156 (22)

*Figures in parentheses indicate approximate duration in weeks.

Table 11.3: Strains of Lac Insect Associated with different Hosts in different Seasons and their Duration of Maturity

Infra Sub Specific Forms (strains)	*Crop*	*Inoculation of Brood Lac*	*Maturity (Harvesting)*	*Duration in Months*
Rangeeni	Rainy season (Katki)	June-July	October-November	4
	Summer (Baisakhi)	October-November	June-July	8
Kusumi	Summer (Jethwi)	January-February	June-July	6
	Winter (Aghani)	June-July	January-February	6

Results and Discussion

From the study, it is revealed that three species of host plants palas such as (*Butea monosperma*), ber (*Zizyphus mauritiana*), kusum (*Schleichera oleosa*) are suitable for commercial lac cultivation which is more or less similar to the findings of (Roonwal, 1962; Kumar *et al.,* 2002: Pal. 2009: Singh *et al.,* 2009: Mohanta *et al.,* 2014; Shah *et al.,* 2018).

A number of major and minor predators as well as Parasitoids usually destroy the lac crop not only by reducing the yield drastically but also by affecting the quality of the lac resin. The vertebrates like monkey, squirrels, rats, lizards, bats, woodpeckers, termites and other birds in the lac ecosystem also pose problem in lac production, although they are not directly involved in either predating or parasitizing the lac insect. The predators account major damage (up to 35 per cent) of lac crop. The following are the predators recorded in the study sites those are associated with the lac insect.

These predator's moths lay their eggs on the infested lac twigs and branches. After hatching, their larvae feed on the lac insects and also on lac encrustations. The white lac moth causes more destruction on trees; while black lac moth, on the stored lac. In addition to these insect pests, squirrels and monkeys also damage lac crops. The rodents greatly damaged the crop. These pests gnawed the mature lac encrustations on the trees or brood lac sticks tied for inoculation and consumed gravid females. The brood lac can be made to fall on the ground by these animals' preventing inoculation.

During the course of study three types of predators (Table 11.4) in addition to four types of Parasitoids and Hyper-parasitoids of each were found on lac crops. Among the predators *Pseudohypatopa* (= *Holcocora*) *pulverea* Meyr (black lac moth) was most dominant species followed by *Eublemma amabilis* Moore (white lac moth) and *Chrysopa spp.* (lacewing bug). It was observed that in the first instar of lac insects the larvae of *Chrysopa spp.* (green lacewing bug) started feeding to the tiny lac crawlers. *Eublemma amabilis* Moore (white lac moth) appearance continued unto 10 days, while for *Pseudohypatopa* (=*Holcocora*) *pulverea* Meyr (black lac moth) it was up to 67 days from the same sticklac. Bhattacharya (2007) recorded prevalence of 14 enemy insects among which *Eublemma amabilis* and *Pseudohypatopa* (= *Holcocora*) *pulverea* are dominant predators. The present study agrees with the findings of Bhattacharya *et al.* (2007); Bhattacharya *et al.* (2004); Jaiswal *et al.* (2007) and stated that *E. amabilis* Moore (Lepidoptera: Noctuidae) and *P. pulverea* Meyr (Lepidoptera: Blastobasidae) are the two dominant predators of lac insect that bring lac crop loss varies from 20 to 40 per cent.Recorded minor Hyper-parasitoids (beneficial fauna) were *Brachymeria tachardiae, Pristomerus sulci, Bracon greeni* Ashmead and *Apanteles spp.* were in rangeeni (katki) and kusmi (aghani) strain. More or less similar findings were made by Jaiswal *et al.* (1998; 2001). As Jaiswal *et al.* (1998) ascertained that the lac insect *Kerria lacca* (Kerr.) associated predators and parasitoids under storage condition. Jaiswal *et al.* (2001) also revealed the lac

Table 11.4: Insect Fauna Associated with Lac Insect *Kerria lacca* (Kerr.) in Rangeeni Strain during 2017-18

Sl. No.	*Associated Fauna Name*	*Order*	*Family*	*Status*	*Range (Mean number of insects per insect per 30 cm stick lac)*	
					August 2017 to October 2017	*November 2017 to July 2018*
	Predators					
1	*Eublemma amabilis* Moore (white lac moth)	Lepidoptera	Noctuidae	Major Predator	0.89-5.65	0.95-5.53
2	*Pseudohypatopa* (= *Holcocora*)*pulverea* Meyr (black lac moth)	Lepidoptera	Blastobasidae	Major Predator	1.38-7.98	1.23-8.40
3	*Chrysopa spp.* (lacewing bug)	Neuroptera	Chrysopidae	Major Predator	1.00-3.85	1.20-3.98
	Parasitoids					
1	*Tachardiaephagus tachardiae* Howard	Hymenoptera	Encyrtidae	Major	2.22-7.43	2.20-6.00
2	*Eupelmus tachardiae* Howard	Hymenoptera	Eupelmidae	Minor	0.19-0.90	0.16-0.90
3	*Parechthrodryinus clavicornis* Cameron	Hymenoptera	Encyrtidae	Minor	0.10-1.25	0.23-1.30
4	*Aprostocetus purpureus*	Hymenoptera	Eulophidae	Major	0.70-3.55	0.85-3.25
	Hyperparasitoids					
1	*Brachymeria tachardiae*	Hymenoptera	Chalcidae	Minor	0.33-1.21	0.35-1.05
2	*Pristomerus sulci*	Hymenoptera	Ichneumonidae	Minor	0.35-0.65	0.20-0.55
3	*Bracon greeni* Ashmead	Hymenoptera	Braconidae	Minor	0.15-1.63	0.19-1.35
4	*Apanteles spp.*	Hymenoptera	Braconidae	Minor	0.18-0.88	0.11-0.54

insect associated predators and parasitoids in lac growing areas of Odisha. These observations similar to species of insect- predators and parasitoids as reported in conventional lac growing areas of north and eastern Indian (Srivastava *et al.*, 1976; Jaiswal *et al.*, 2001 a; b). The present study is also more or less similar to Meena *et al.* (2018) who reported that 11 species of fauna associated with *Kerria lacca* from 8 families under 3 were recorded representing predator species *E. amabilis, P. pulverea, C zastrowi; primary parasitoids T. tachardiae, A. purpureus, T. clavicornis, E. dewitzi and hyper-parasitoids A. fakhrulhajiae, E. tachardiae, B. greeni, B. Tachardiae* in western plains of India.

Conclusion

Among the different lac host plants Kusum (*S. oleosa*) was found best host plant in kusmi strain for commercial lac production. Three major predators namely *Pseudohypatopa* (= *Holcocora*) *pulverea*, *Eublemma amabilis* and *Chrysopa spp.* are three major predators belonging to three families and two orders. Among the parasitiods *Tachardiaephagus tachardiae* and *Aprostocetus purpureus* as major parasitoids whereas *Eupelmus tachardiae* and *Parechthrodryinus clavicornis* was recorded as minor parasitoids belonging to three families and one order. Whereas hyper parasitoids of four species three families and one order is found in the periphery of Kuldiha Wildlife Sanctuary of Odisha.

REFERENCES

Ahmad A, Kaushik S, Ramamurthy VV, Lakhanpaul S, Ramani R, Sharma KK and Vidyarthi AS (2012) Mouthparts and stylet penetration of the lac insect *Kerria lacca* (Kerr) (Hemiptera: Tachardiidae). Arthropod Struct Dev 41: 435–441.

Ajaz-ul-Islam M, Quli SM, Rai R and Sofi PA (2013) Livelihood Contributions of Forest Resources to the Tribal Communities of Jharkhand. Indian Journal of Fundamental and Applied Life Sciences (ISSN: 2231-6345).

Bharat R (2010) Current status of lac production issue, remedial measures and support system for development. District Forest Officer, Jharkhand Forest Department, Palamau, Jharkhand, pp 33-34.

Bhattacharya A, Jaiswal AK, Mishra YD and Chakrabarty PK (2004) Efficacy of *Trichogramma brasiliense* Ashmead, in the management of Lepidoptera predators of lac insect, *Kerria lacca* (Kerr) Lac production Division, Indian lac Research Institute, Namkum, Ranchi Jharkhand.

Bhattacharya A, Kumar S and Jaiswal AK (2007) Evaluation of Trichogramma species for the suppression of lepidopteran insect predator, *Eublemma amabilis* Moore, in lac culture on Flemingia macrophylla. J Bio Cont 21(2): 267-270.

Chattopadyay S (2011). Introduction to lac and lac culture, Bulletin, Department of forest biology and tree improvement faculty of forestry; Birsa Agriculture University, Kanke, Ranchi, Jharkhand.

Chen XM (2005) Biodiversity of Lac Insects. Yunnan Science and Technology Press.

Chowdhury MK, Sen P and Battacharya A (1971) Studies on the relationship between the density of lac insect, *Kerria lacca* (Kerr.) (Homoptera: Coccidae) on chalcid parasitization. Indian Journal of Entomology 33: 180-182.

Das R and Kumar A (2013) Lac cultivation and rural livelihood. In: Kumar A, Das R (eds) Prospects of scientific lac cultivation in India. IFP, pp 217–231.

Eben A, Benrey B, Sivinski J and Aluja M (2000) Host Species and Host Plant Effects on Preference and Performance of *Diachasmimorpha longicaudata* (Hymenoptera: Braconidae). Environmental Entomology 29: 87-94.

Ghosal S (2013) Management of host pants for lac cultivation. In: Kumar A, Das R (eds) Prospects of scientific lac cultivation in India. IFP, pp 48–58.

Ghosh J, Lohot VD, Singhal V, Ghosal S, Sharma KK and Ramani R (2014). Plant-insect-environment interaction for kusmi lac production in ber (*Ziziphus mauritiana*) varieties, *The Eco scan*, Special issue, Vol. VI: 407-411.

Ghosh J, Lohot VD, Singhal V, Ghosal S and Sharma KK (2016) Morphological-biochemical-physiological traits assisted selection for kusmi lac production on ber (*Ziziphus mauritiana* Lam.) varieties. Indian J. Hort. 73(1): 19-24.

Ghugal SG, Thomas M and Pachori R (2015) Performance of Katki Lac on Nutrient Managed of *Butea monosperma* (Lam.) Taub. Trends in Biosciences 8(24): 6873- 6877.

IINRG (2018) Lac, plant resins and gums statistics: at a glance. Bulletin (Technical) No. 19/2018. ICAR-IINRG, Ranchi.

Jaiswal AK, Bhattacharya A, Kumar S and Patamajhi P (2007) Evaluation of ethofenprox on *Chrysopa madestes* (Neuroptera: Chrysopidae) a serious pest of Indian lac insect, *Kerria lacca* Kerr. J Ento Res 31(2): 113- 114.

Jaiswal AK, Bhattacharya A, Sushil SN and Kumar P (2001a) Incidence of lac associated insect fauna in a few lac growing area of Jharkhand. Indian Journal of Applied Entomology 15: 55-57.

Jaiswal AK, Bhattacharya A, Sushil SN and Kumar P (2001b) Incidence of lac associated insect fauna in a few lac growing area of Orissa. Journal of Applied Zoological Research 12: 75-77.

Jaiswal AK, Pal G, Singh JP and Patel B (2012) Lac production growth analysis for the state of Odisha. Bio-ved 23(1): 1-5.

Jaiswal AK, Sharma KK and Agarwal SC (1998) An efficient and indigenous device for lac-insect pest management. Tropical Science 38: 81-86.

Jaiswal AK, Sharma KK and Kumar KK (2006) Importance of Lac in the socio-economic life of tribals in Ranchi District (Jharkhand). J. Non-Timb Forest Prod 13(1): 47-50.

Jaiswal AK, Sharma KK and Kumar KK (2006) Importance of lac in the socio-economic life of tribals in Ranchi district (Jharkhand). New Agric 17(1/2): 133–137.

Jaiswal AK and Sharma KK (1998) An efficient and indigenous device for lac insect pest management. Tropical science 38(2): 81-86.

Jaiswal AK, Bhattacharya A, Sushil SN and Kumar P (2001) Incidence of lac associated insect fauna in certain lac growing areas of Odisha. J. Appl. Zool. Res. 12(1): 75-77.

Janghel S (2013) Study on comparative efficacy of insecticides in Katki crop for predator management on Rangeeni lac crop on *Zizyphus mauritiana* in Malara village, Seoni District. M.Sc. Thesis. JNKVV, Jabalpur, M.P.

Kerr J (1782) The philosophical transaction. Royal Soc of London 71: 374-382.

Khobragade D, Thomas M, Pachori R, Sharma JL and Shrivastava A (2012) Farmer participatory trial on the predator management of Lac insect *K. lacca* (Kerr.) in Anuppur District, Madhya Pradesh. Journal of Tropical Forestry, 28(1and2): 38-45.

Krishnaswami S (1962) Lac through the ages, In Mukhopadhayay B., Muthana M. S. (eds.), A monograph on lac. Indian Lac Research Institute;Ranchi, India, pp. 1–13.

Kumar A and Ramani R (2010) Current issues related to Lac production, Compilation of talks, 20-21 September, 2010, pp 56.

Kumar KK, Ramani R and Sharma KK (2002) Scope of lac cultivation in employment and income generation. In: Recent advances in lac culture, ILRI, Ranchi., 254-262.

Meena SC and Sharma KK (2018) Species composition and richness of entomo-fauna associated with lac insect, *Kerria lacca*, Kerr in western plains of India, Journal of Entomology and Zoology Studies 6(3): 684-690.

Mohanasundaram A, Monobrullah Md, Sharma KK, Meena SC, Verma S and Ramani R (2016) Lac insect and associated fauna-*A Practical manual*. ICAR-Indian Institute of Natural Resins and Gums, Ranchi (Jhharkhand), India. Bulletin (Technical) No. 2016 14: 01-42.

Mohanasundaram A, Monobrullah Md, Sharma KK, Meena SC, Verma S and Ramani R (2016) *Aprostocetus purpureus*, a major parasitoid of Indian lac insect, *Kerria lacca* (Coccoidea: Tachardiidae). Indian Journal of Ecology 43: 517-521.

Mohansundaram A, Meena SC, Monobrullah Md, Sharma KK, Anees K and Verma S (2018) Variations in population of lac associated fauna in relation to different lac host plants, lac insect strain and seasons of cultivation, Journal of Entomology and Zoology Studies,6(02): 490-495.

Mohanta J, Dey DG and Mohanty N (2012) Performance of lac insect, *Kerria lacca* (Kerr.) in conventional and non-conventional cultivation around Similipal Biosphere Reserve, Odisha, India. Bioscan 7: 237-240.

Mohanta J, Dey DG and Mohanty, N (2014) Studies on lac insect (*Kerria lacca*) for conservation of biodiversity in Similipal Biosphere Reserve, Odisha, India. Journal of Entomology and Zoology Studies, 2(1): 1-5.

Monobrullah Md, Mohanasundaram A, Meena SC, Sweta V and Sharma KK (2016) Host and location mediated variation in life cycle and biological attributes of Indian lac insect, *Kerria lacca* (Kerr.) Indian Journal of Ecology 1: 169-172.

Mulugeta L, Tarekeng A and Olsson M (2003) Gum and resin resources from some acacia, *Boswellia* and *Commiphora* species and their economic contribution in Liban, South eastern Ethiopia. Journal of Arid Environment 55: 465-482.

Namdev BK, Thomas M, Kurmi A, Thakur AS and Upadhyaya A (2015) Impact of nutrient management of *Zizyphus mauritiana* (lamb.) On the yield of kusmi lac. The Bioscan. 10(3): 1219-1222.

Ogle A, Thomas M and Tiwari LM (2006) Strategic development of lac in Madhya Pradesh. Final Report, Department for International Development (DFID), MPRLP-TCPSU, India, June 2006, 1-34.

Pal G, Jaiswal AK and Bhattacharya A (2010) Estimation of lac production and processing in India. Environment and Ecology 28 (1B): 572-576.

Pal G (2009) Impact of scientific lac cultivation training on lac economy- A study in Jharkhand. Agricultural Economics Research Review 22: 139-143.

Prasad R (2011) Strategy for Sustainable NTFP Management in India. International Conference on NWFP for Sustained Livelihood Bhopal- India.

Ramani R, Baboo B and Goswami DN (2007) Lac- An introduction. Indian Lac Research Institute, Ranchi 12.

Ramani R, Sharma KK and Kumar P (2008) A new record of occurrence of Indian lac insect, *Kerria lacca* (Kerr.) (Coccoidea: Tachardiidae) on *Euphorbia pulcherrima* Willd. (Euphorbiaceae). Indian-Journal-of-Forestry. Dehra Dun, India 31(2): 283-284.

Rathore V (2011) Comparative performance of three local Rangeeni Lac cultivar on *Zyziphus maurtiana* in Dharna village, Seoni district, Madhya Pradesh. M.sc. (Ag) Thesis, submitted, JNKVV, Jabalpur.

Roonwal ML (1962) Lac Hosts. In: A Monograph on Lac, Mukhopadhyay, B. and M.S. Muthana (Eds.). ILRI, Ranchi, India, 14-58.

Sahu S (2016) Survival and Yield of Rangeeni Lac insect on *Butea monosperma* (Lam) treated with different Micronutrients and Humic acid. M.sc. (Ag) Thesis, submitted, JNKVV, Jabalpur.

Sarvade S, Panse RK, Rajak SK and Upadhyay VB (2018) Impact of biotic and abiotic factors on lac production and people's livelihood improvement in India-An overview. Journal of Applied and Natural Science 10(3): 894- 904.

Shah AI (2016) Importance of Lac culture on socio-economic and biodiversity conservation fronts. Int J Adv Res 4(9): 1708–1711.

Shah TH, Thomas M and Bhandari R (2015) Economic analysis of kusmi lac production on *Zizyphus mauritiana* (Lamb.) under different fertilizer treatments. Journal of Environmental and Agricultural Science 4: 37-41.

Shah TH and Thomas M (2018) Survival of kusmi lac insect (*Kerria lacca* Kerr.) on nutrient managed *Zizyphus mauritiana*. Indian Journal of Entomology 80(1): 56-63.

Sharma and Ramani (eds.) (2011). Recent Advances in Lac Culture, IINRG, Ranchi.

Sharma H, Ghugal SG, Thomas M and Pachori R (2015) Impact of Nutrient Management in *Butea Monosperma* (Lam.) Taub. on the Survivability of *Kerria lacca* (Kerr). Trends in Biosciences 8(23): 6682-6687.

Sharma KK, Jaiswal AK and Kumar KK (2006) Role of lac culture in biodiversity conservation: issues at stake and conservation strategy. Curr Sci 91 (7) : 894-898.

Sharma KK and Ramani R (2010) Genetic variability in lac insects. In: Recent advances in lac culture (eds. Sharma, KK and Ramani R), IINRG, Ranchi, pp. 46-51.

Sharma KK, Jaiswal AK, Bhattacharya A, Mishra YD and Sushil SN (1997) Emergence profile and relative abundance of parasitoids associated with Indian lac Insect, *Kerria lacca* (Kerr.). *Indian* Journal of Ecology 24(1): 17-22.

Sharma KK, Jaiswal AK and Kumar KK (2006) Role of lac culture in biodiversity conservation: issue at stake and conservation strategy. Current Science 91(7): 894–898.

Sharma R (2013) Lac production and industrial economics in India. In: Kumar A, Das R (eds) Prospects of scientific lac cultivation in India. IFP, pp 271–275.

Shiva MP (1993) Solution to overcome impediments in forest development through MFP based management. Paper presented at the international seminar on minor forest products in forestry, 17-18th April, Dehra Dun.

Singh SK (2005) An appraisal of livelihoods in rural Jharkhand. Praxis-Institute for participatory practices Ranchi Jharkhand. Indian Journal of Fundamental and Applied Life Sciences.

Singh JP, Jaiswal AK, Monobrullah Md and Bhattacharya A (2009) Response of some selected insecticides on neuropteran predator (*Chiysopa lacciperdci*) of lac insect Indian J. Agric. Sci., 79(9): 727-731.

Srivastava DC, Mehara BP and Teotia TPS (1976) On the related abundance of inimical and beneficial insects associated with the Indian lac insect, *Kerria lacca* (Kerr). Indian Journal of Ecology 3: 194-196.

Srivastava DC, Chauhan NS and Teotia TPS (1984) Seasonal abundance of insects associated with the Indian lac insect, *Kerna lacca*. Indian Journal of Ecology 11(1): 37- 42.

Srivastava SK and Jeet P (2020) Effect of Irrigation Methods and Mulching on Growth and Seed Yield of Semialata. Journal of Agri Search 7(1): 18-20.

Uiky BK (2015) Effect of Rynaxypyr and Emamectin benzoate on incidence of natural enemies, productivity of lac insect *Kerria lacca* Kerr and residual toxicity against parasitoids. M.Sc. (Ag.) Thesis, IGKV, Raipur, 30-55.

Varshney RK (1976) Taxonomic studies on lac insects of India (Homoptera: Tachardiidae). Orient Insects (Supplement) 5: 1-97.

Yogi RK, Kumar A and Singh AK (2018) Lac, plant resins and gums statistics 2018: At a glance, Bulletin (Technical) No. 19/2018, (ICAR-Indian Institute of Natural Resins and Gums, Ranchi, India),1-80.

Yogi RK, Kumar A and Jaiswal. Lac, Plant Resin and Gum Statistics (2015) At a Glance. ICAR-Indian Institute of Natural Resins and Gums (IINRG), 2015: 1-72.

Chapter 12

Prevalent Agroforestry Systems of Eastern India: A Viable Option of Land Management System for Livelihood Security and Sustainability

G. Shial[1], K. Upadhyaya[2], S. Nayak[3] and P. Pallavi[4]

[1]*Krishi Vigyan Kendra, Bhadrak, OUAT–756111, Odisha*
[2]*Department of Forestry, Mizoram University–796004*
[3]*College of Forestry, OUAT–751003, Odisha*
[4]*College of Agriculture, OUAT, Bhawanipatna–766002, Odisha*
e-mail: gayadharbalasore@gmail.com

Introduction

Agroforestry system of land use is rapidly gaining importance in the recent years for meeting the dual objectives of achieving food security and environmental sustainability. It not only improves the farm production and income but also conserves the natural resources and ensures their optimum utilization. This type of land use has been instrumental in reducing the CO_2 emissions from agro-ecosystem, stabilizes the atmospheric CO_2 concentration and hence minimizes the impact of climate change. In recent years more attention has been given to agroforestry for sustainable land use system particularly in the tropics. Deliberately rearing of woody components on the farm land is an age old practice of farmers to get food, fodder, timber, fuelwood and to obtain shade, shelter, combat erosion, soil fertility improvement and to improve avenue/aesthetic benefits of landscape (Cittapur and Patil, 2017).These traditional tee based farming system has been gradually disappearing due to low economic values as compared to conventional agricultural practices (Nerlich *et al.,* 2013).The low economic significance of these traditional land use system may be due to lack of adequate sustainable management of agroforestry systems. Moreover, it has been found that the practice of conventional agriculture resulted in permanent changes in function and structure of ecosystem. In India, per capita arable land and forest land availability is less than 0.2 ha and 0.1 ha respectively, due to which there is extreme shortage of timber, fuel wood and fodder. Further rapid increasing of human and livestock population, industrialization and urbanization promote land degradation and limiting sustainable production. There is no availability of additional land for horizontal expansion of agriculture. So, need of the hour is to find viable technologies for improving productivity and resource conservation on sustainable basis. Further, small and marginal farmers, who constitute 84 per cent of the farming community in the country are mostly suffers due to natural calamities like drought, flood, cyclone and heat wave and unprecedented market behavior have encouraged changing agricultural land for non-farm activities. In the era of shrinking resource base of land,water and energy, resource use efficiency is an important aspect for considering the sustainability of crop production. There are wide availability of unutilized uplands in the country to be properly utilized for enhancing rural economy of the farmers. In this context, agroforestry system offers vast potential to counteract the negative impacts of intensive agricultural practices that diversifies and sustain overall land productivity for increased socio-economic and environmental benefits (Verchote *et al.,* 2007, Dhillon *et al.,* 2010). It meets the demand of fuel wood, fodder, timber, medicine, other non-timber forest products (NTFPs),enhanced green cover, soil and moisture conservation and reduce the pressure on forest. In India, out of 329 M ha hectare of geographical area, 145 M ha is cultivable, with per capita cultivated land availability coming down from 0.48 ha in 1952 to less than 0.15 ha at present posing a threat to the security of food and livelihood.

Alternate upland management system is suitable for appropriate utilization of uplands. It refers to bringing about a desirable change in the existing cropping system/pattern towards more balanced ones to meet the ever-increasing demand of population. Diversification of cropping system is necessary to get higher yield and return, to maintain soil health, preserve environment and meet daily requirement of human and animal population. It also aims at employment generation, profit generation with proper utilization of available resources. The uplands of our country offer vast scope for crop diversification with multiple options of growing tree crops. Adoption of tree based intercropping systems stabilize the economy of rainfed farmers in aberrant weather years. These may be based on the available resources and existing climatic situations with an objective to increase profit per unit land. Among various states of Eastern India (Eastern Uttar Pradesh, Bihar,

Jharkhand, Odisha, Chhattisgarh, West Bengal and Asam), Jharkhand has the highest area under wastelands/degraded lands (14.84 per cent), followed by Assam (11.20 per cent) and Odisha (10.69 per cent) (World Bank,2007). There is an urgent need to improve those wastelands/degraded lands through agroforestry interventions in order to meet the increasing demands of food, fuel, fodder and timber requirements. Thus agroforestry acts as one of the option to solve the problems of resource overexploitation and it's degradations in this region.

Agroforestry: The Viable Alternate Upland Management

Agroforestry is a collective name of upland management systems where woody perennials(trees, shrubs, palms, bamboos *etc.*.) are deliberately used on the same land management units as annual agricultural crops and/animals in some form of spatial arrangement or temporal sequence. In agroforestry system there is both ecological and economic interactions between different components (Lundgren and Raintree,1982). Tree farming, wood lots, pasture and grass land systems, ley farming and agroforestry systems are different forms of the viable alternate upland management (Singh,1997; Singh and Osman, 1995).It provides assured income and minimize risk through efficient utilization of existing resources and adequate technology for rehabilitating degraded marginal/wastelands and protecting cultivable lands from further degradation. In vagaries of climatic conditions these systems can withstand and utilizes the resources more efficiently. Complementary relationships between trees, crops and livestock components brings sustainability, productivity, stability and profitability of the system. It facilitates nutrient recycling, improves the soil fertility, arrests soil erosion and land degradation, conserves soil and water, improves village environment, conserves biodiversity and serves as a major carbon sink. Realizing the importance of this type of land use, emphasis is being given to promote agroforestry in agricultural lands and permanent fallow lands for stabilizing food production and extending the environmental services. But most of the agroforestry systems are part of indigenous traditional knowledge of local communities. These systems vary from one part of the country to another due its diverse climatic conditions. In recent past population of India is increasing at an alarming rate, which hinder the limited availability of cultivated land and natural resources of country. The land-holding size of farmers gradually shrink and agroforestry is the only way to optimize the farm productivity (National Agroforestry Policy, 2014).Gross productivity under the agroforestry systems is greater as compared to mono cropping and farmers get better financial return from this integrated farming. A detailed study shows that economic viability with internal rate of return ranging from 25 to 68 and B:C ratio of 1.01 to 4.17 for 24 agroforestry systems from different agro climatic regions of country (Chavan *et al.*, 2015). In India more than 75 per cent population depends on agriculture for survival and income (Thenmozhi *et al.*, 2014). Land not suitable for agricultural crops due to land degradation, salt affected soil, shallow soil condition and anthropogenic interference should have brought under agroforestry activities.

Traditional Agroforestry Systems in India

Traditional Agroforestry Systems in India are mainly based on existing indigenous knowledge and tree species are selected as a part of age-old cultural practices as well as cultural pattern of the farming community. They integrate varieties of woody plants in the crop and animal production systems depending upon the agroclimatic conditions and local needs of the farmers (Dhyani, 2011). In India traditional agroforestry system varies from simple forms of shifting cultivation to complex home gardens, representing sparse stands of trees to high density multistoried system respectively. This provides the basic needs such as fuel wood, fodder, timber and maintains the livelihood of the rural people (Kareemulla *et al.*, 2009, Bijalwan *et al.*, 2011, Jamala *et al.*, 2013).They use fuel wood, tree borne oil seeds (TBOS) as a major source of heating and cooking, timber for construction, furniture, agricultural implements and fodder requirements for livestock particularly during summer season. Before the evolution of agroforestry as a concept and technology, farmers have been maintaining a variety of tree species in their backyard home garden and boundary of crop fields along with their agricultural crops since time immemorial. It has been evolved along with the development of human civilization and continues to play a vital role in maintaining the sustainability of rural livelihoods. The indigenous agroforestry systems such as home gardens, scattered trees on farm lands or field bunds and fodder trees for green fencing are practiced by farming communities These traditional agroforestry systems are deeply linked with the livelihoods of the resource poor small holders, offer a wide range of goods and services and insulate them from the climatic uncertainties. Besides, it provides a valuable niche for the diverse flora, fauna and micro-organisms including the pollinators, decomposers, predators and nutrient solubilizes. These agroforestry systems are specific to agro-climatic zone, agro-ecological situation and social set up of the community. Therefore, diverse tree species are grown and maintained traditionally by the community of different regions of the country for deriving their multiple needs. In Odisha diverse traditional agroforestry systems exist in different areas characterized with high species diversity including multipurpose tree species, shrubs and palms. In the drought prone western parts *Acacia nilotica and Butea monosperma* based farm forestry is the dominant one and in the central inland districts *Madhuca latifolia* and *Scleichera oleosa* based agroforestry are more prevalent. However, in the coastal areas of Odisha tree species like Palms, *Albizia lebek, Gmelina arborea, Pongamia pinnata, Erythrina indica, Samanea saman, Terminalia spp, Bambusa strictus etc.* are maintained by the farmers along with their agricultural crops in their backyard as home garden or in the field as farm forestry. Agroforestry in the coastal areas has the potential to improve the livelihoods of vulnerable rural communities by supplementing their household income and reducing their exposure to climate risks. The traditionally palm based (coconut, *Cocos nucifera* and palm, *Borassus flabellifer*) agroforestry are still prevalent in the coastal area for food during lean season, fire wood and building materials. Generally, the backyard home gardens are multi-tier systems characterized by a high density and diversity of plant species consisting of trees/palms, shrubs, climbers and seasonal crops that meet the multifarious need of the rural households from small piece of land. Restoration of water logged and salinity affected areas of eastern India brought by the agroforestry interventions of suitable plant species act as a bio drainage system. The excess soil water gets drained out by deep rooted plants through the process so called bio-energy (Chauhan *et al.,* 2012).Fast growing plants with deep tap root systems are selected for effective management of such areas. Many fast-growing species like *Dalbergia sissoo, Eucalyptus spp., Pongamia pinnata, Syzygium cumini, Terminalia arjuna etc.*. are found to grow in such water-logged conditions. Lac cultivation in agrisilvicultural system is an age-old practice of the North Eastern states of Jharkhand, Chhatisgarh *etc.*. Plant species like *Schleichera oleosa* (Kusum), *Zizyphus mauriatiana* (Ber), *Butea monosperma* (Palas) and *Flemingia semialata* (Bara solpan) are most suitable for lac cultivation in agrisilvi system in field bunds or in scattered patches in and around the hamlets.

However, over the years these traditional agroforestry systems are rapidly disappearing in all the eco-systems in general and coastal environment in particular due to high population pressure and subsequent land fragmentation and diversion of land for habitat purpose. Many of the plant species are rapidly losing their significance and are in the verge of extinct making the ecology of eastern India more vulnerable to climate extremes. Traditional agroforestry systems are found in Indian condition (Pathak *et al.,* 2000).Some prominent diversity of Agroforestry systems are the Shifting cultivation system of the North east, multi-storey home gardens of Kerala and Tamil Nadu, Taungya system of Uttar Pradesh. The traditional agroforestry systems with diverse features are listed in Table 12.1.

Table 12.1: Traditional Agroforestry Systems Prevalent in India

Sl. No.	*Traditional Agroforestry Systems*	*Agro-ecological Adaptations*	*Features/ Combinations*	*Reference*
1	Shifting cultivation	In tropical forest areas. Mainly North east states and Tribe's supported areas.	Cyclic slash and burn, growing of arable crops, fallow and back; public and private lands. patch of forest cleared and burn then Grain crops are grown on those blank patches for some years after this the same patch will leave fallow for regeneration of vegetation.	Garbyal (1999)

Sl. No.	Traditional Agroforestry Systems	Agro-ecological Adaptations	Features/ Combinations	Reference
2	Taungya	In all regions but in large scale in U.P., U.K., Cachar regions of Assam and hilly areas where land holding is very minute in quantity. Woody plant+ Arable crops. Main purpose to reduce weed growth in early stage of plantation (Departmental Taungya), for sustainable livelihood and income generation (In village Taungya and Leased based Taungya).	Tree planting, Intercropping upto 3 years, moving to other places; Govt. land and trees, intercrops for cultivators; Sharing of short-term products between state and landless farmers.	Rowbotham (1924)
3	Home gardens	Mainly in humid tropics of Southern part of country where rain fall is above 2500mm	Transitory stage between tropical forest ecosystem and arable cropping; Plantation, fruit,spices,vegetable, fuel, fodder and MPT in backyard of multiple strata; Domestic needs and biodiversity conservation.	Kumar (2010)
4	Large Cardamom and Alder plantation	Mainly humid tropics	Cardamom supported by Alder and other spices; Tree harvesting at 15cm dia. Lopping after cardamom maturation and pollarding at 2.0 – 2.5 m. height for regeneration of younger seedlings and open canopy to regulate light; Soil fertility, shade, fodder, employment for labour and better income.	

Sl. No.	*Traditional Agroforestry Systems*	*Agro-ecological Adaptations*	*Features/ Combinations*	*Reference*
5	Plantation or cash crop based orchards	(i) All regions	Intercropped with cereals and other field crops; Short term fruits; Tubers and fodder crops; Enhanced resource use efficiency; Employment, livelihood support and income.	
		(ii) Plantations like coffee + fruit trees/ ornamental tree + spices/vegetables in humid and sub-humid tropical regions	Mixing of tree with different strata with plantation crops.	Dagar (2014)
		(iii) Tea plantations + shade trees North East	Plantation crops+ Shade tree	
		(iv) Punjab, Haryana, U.P *etc..*	Fast growing tree species of industrial value+ Field crop	Chauhan *et al.* (2012)
		(v)Very high slope where arable crop cannot grow	Fruit Tree + Grass land	Singh (2008)
		(vi) Coastal region of country	Coconut based mix farming.	Dagar (1991, 1995 and 2000)
6	Scattered Trees on outside or inside the farm Lands	All regions specially in arid and semi-arid on the field of small and marginal farmers. MPTs trees scattered on farm either in the form of wind break, shelter belt or scattered in the field.	Naturally occurring or planted timber,fruit, MPTs and main field crops; Income,wood supply and some fodder in critical period.	Tewari (2008)

Sl. No.	*Traditional Agroforestry Systems*	*Agro-ecological Adaptations*	*Features/ Combinations*	*Reference*
7	Shelterbelts and wind breaks	In wind prone arid, alpine areas and coastal regions. In all regions but mainly where wind speed is very high and soil erosion marked like Rajasthan..	Plantation against sea tides, hot winds *etc.*; Income; Environmental protection and domestic needs. Crops are grow on leeward side of tree component to reduce the wind speed.	Dagar *et al.* (2014)
8	Boundary plantations and Live Hedges. Cut and carry systems improved form of silvo-pasture system also called protein banks.	All regions. Research farm, dry and waste land.	MPTs on field crops and orchard boundaries; Income, Employment; Supply for domestic needs and industry, crop protection and wind break. Leguminous trees are grown along with grass land. Example- *Acacia nilotica*+Grasses (highly palatable)	Mathukia *et al.* (2016)
9	Woodlots for Soil Conservation	In hilly areas, along sea coast and ravine	Resource conservation, ground water recharge, amelioration of microclimate.	
10	Trees on Rangelands	All regions	Shrubs and bushes in community grazing lands; Resilience against climatic aberrations; Green biomass during lean months and higher production in silvi-pastures.	
11	Plantation crops with pastures and animals	Mostly in humid and sub-humid regions particularly in hills and dry region where rainfall is scanty and soil depth is very less.	Green biomass during lean months and higher production in silvi-pastures.	Singh (2008)

Sl. No.	*Traditional Agroforestry Systems*	*Agro-ecological Adaptations*	*Features/ Combinations*	*Reference*
12	Industrial Plantations with Crops	Intensively cropped areas in North India	For fuel, pole and pulpwood; income, Domestic needs and for industrial uses.	
13	Seasonal Forest Grazing	Semi-arid and mountainous ecosystem	Migratory grazing, Reduced fire hazards, increase soil fertility, livelihood support and income.	
14	Aqua agforestry	Low lying lands. Followed in many places in N.E states like Assam and Bastar region of Chhattisgarh.	Fish pond dikes and surface planted with suitable species; Source of nutrition to aquatic life, breeding ground, nutrient cycling; Shade for crustaceans; Employment and higher income. Aqua agroforestry (*Acacia nilotica*, *Syzygium cuminii* grow on the boundary of fish pond).	Chavan *et al.* (2015)
15	Inter cropping	All region of the country	N2 fixing trees+ Crops/ Grasses	
16	Wadi form of cultivation	Practices are carried out in Gujarat, M.H, Karnataka and some parts of Rajasthan with the help of BAIF.	Arable crop are grown with intercropping of fruit tree inside the field and forestry tree species are grown on the boundary of field.	Hegde (2005)
17	Agri-silvi-apicuture	All regions. Punjab, Haryana, Parts of M.P., Rajasthan *etc.*.	Agriculture crop (mainly mustard) + Tree + Honey bee rearing. Trees as a medium for apiculture; Employment and Income for landless and resource poor; Better pollination of other crops.	Gill *et al.* (2010)

Prominent Agroforestry Systems in India

The rural community in Agroforestry system get fuel wood, fodder, fruits and fibres up to a larger extent in to improve the living conditions of the people through livelihood support and alleviating poverty (Bijalwan *et al.,* 2013).Tree-based systems are more productive, more sustainable and more harmonized to people's cultural or material needs than treeless alternatives (Mbow *et al.,* 2014). The prominent kind of present Agroforestry systems practiced by the farmers of India are classified based on the nature of the component and the pattern of tree planting on the farm land and the number of farmers practicing specific agroforestry systems were recorded and expressed in terms of priority ranking of the system by farmers of. Agroforestry systems differs in different places according to specific relevance of the environmental, socio-cultural needs of inhabitant societies of a region. Present agroforestry practices and major emphasis on agroforestry in different regions of India are presented in Tables 12.2 and 12.3 respectively (Pathak *et al.,* 2006a).

Table 12.2: Prominent Agroforestry Systems in different Agro-climatic Regions of India

Sl. No.	*Agro-Climatic Zones*	*Agroforestry Systems*					
		Agri-silvicul-tural	*Agri-horti-silvicul-tural*	*Agri-horticul-tural*	*Horti-pasto-ral*	*Silvipas-toral*	*Specialized System*
1.	Western Himalayan Region	4	4	5	4	4	-
2.	Eastern Himalayan Region	5	3	5	4	3	Jhum
3.	Lower Gangetic Plains Region	5	-	3	2	2	Energy
4.	Middle Gangetic Plains Region	4	-	2	-	-	Homestead
5.	Upper Gangetic Plains Region	5	-	5	-	-	Agrihorti-silviculture
6.	Trans- Gangetic Plains Region	5	-	5	-	-	Agrihorti-silviculture
7	Eastern Plateau and Hills Region	5	-	4	3	4	Block Plantation
8.	Central Plateau and Hills Region	4	-	5	4	4	Block Plantation
9.	Western Plateau and Hills Region	4	-	5	2	4	-
10.	Southern Plateau and Hills Region	5	-	3	2	4	-

Sl. No.	Agro-Climatic Zones	Agroforestry Systems					
		Agri-silvicultural	*Agri-horti-silvicultural*	*Agri-horticultural*	*Horti-pastoral*	*Silvipastoral*	*Specialized System*
11.	East Coast Plains and Hills Region	4	5	4	2	2	Home garden/ shelter belt
12.	West Coast Plains and Ghats Region	4	5	4	2	2	Home garden
13.	Gujarat Plains and Hills Regions	4	2	4	3	5	Block Plantation
14.	Western Dry Region	5	3	4	-	2	shelter belt
15.	The Islands Regions	4	5	4	1	2	Home garden

*No. 1-5 indicate the priority practiced, model 1= Least and 5 = Highest.

Commercial Agroforestry Practices

It has now been well recognized that agroforestry can address some of the major land-use problems of rain fed farming systems in India and that can be modified by improving the indigenous systems. These practices need to be further improved with suitable technological interventions considering need of the local people, so that socio-economic status of the farming community will be increased. In the present scenario attempts are being made to introduce commercial agroforestry practices using multipurpose tree species.

However, there has been a growing realization of the benefits of tree crop integration systems and since 1980s there is a widespread interest in planting of a new species of Acacia that is native to Australia, *Acacia mangium*, popularly known as Australian teak in the home garden, farm boundaries and permanent fallows. Farmers are also planting and maintaining tree species like *Acacia auriculiformis, Dalbergia sissoo, Swietenia macrophylla, Gmelina arborea etc.*. along with their agricultural crops as a part of commercial agroforestry systems. Agroforestry models adopted by farmers are highly lucrative, thus attracting farmers in a big way. In these areas *Acacia auriculiformis* and *Acacia mangium* planted on agricultural fields and field boundaries is harvested at 10 to 12 years, and the average economic return of *Acacia auriculiformis* and *Acacia mangium* based agroforestry systems are high compared to that of sole agriculture crop. Poplar tree (*Populus spp.*), *Eucalyptus spp.* are also popularized among farmers of different states of our country. In spite of multiple benefits from agroforestry practices, majority of farmers have been hesitant to adopt these systems on large scale primarily because of certain apprehensions about the tree component such as long rotation, reduction in gross

Table 12.4: Main Characteristics and Agroforestry Emphasis in the Major Agro Ecological Regions of India

Characteristics	*Himalayan Region*	*Indo-Gangetic Alluvial Plains*	*Arid and Semi-Arid Region*	*Humid and Sub-Humid Region*	*Tropical Coastal and Island Region*
Climate	Varies from temperature cold alpine to sub-tropical; rainfall 8 to 350 cm in west, 200-400 cm in east	Very hot summer and cold winter; rainfall 30 to 200 cm: ground frost common in winter	Dry-humid to very hot, sub-humid; rainfall in arid 40-65 cm; in semi- arid 70-125 cm	Humid and sub-humid; hot: rainfall 100 to 400 cm	Tropical humid, lowlands; rainfall 60-310 cm
Geographical spread	Uttaranchal, Himachal Pradesh, Jammu and Kashmir; elevation 1,000-3,000 m	Below foot hills of Himalaya covering about 47 M ha of Punjab, Haryana, Delhi, Jharkhand, Uttar Pradesh, Bihar, Parts of West Bengal; elevation 150 to 600 m	Spreads over Rajasthan, Gujarat, Punjab, Haryana; parts of Uttar Pradesh, Madhya Pradesh, Andra Pradesh, Maharashtra, Karnataka and Tamil Nadu	Spreads over parts of Assam, Mizoram, Tripura; West Bengal, Odisha, Madhya Pradesh, Chhattisgarh and the Southern states, elevation 150 to 1,500 m	Coastal regions of West Bengal, Odisha, Andra Pradesh, Tamil Nadu, Kerala, Karnataka, Maharashtra and the Islands
Soils and vegetation	Light to heavy textured, mostly Inceptisols. Alpine scrub sub-alpine coniferous, tropical evergreen, semi-evergreen forests, alpine grasslands	Mostly alluvial with patches of saline and alkali soils. Plant woodlots of *Eucalyptus spp., Populous spp., Acacia spp. etc.*. xerophytic trees and shrubs occur naturally in degraded soils	Sandy to clay loam; xerophytic thorny scrub forests and woodlands and psamophytic scrub vegetation, grazing lands	Alluvial to clay loam; Oxisols, Alfisol Evergreen and deciduous forests	Entisols, Oxisols, Inceptisols. Alfisols. Evergreen, semi-evergreen, moist deciduous, littoral and mangrove forest

Characteristics	*Himalayan Region*	*Indo-Gangetic Alluvial Plains*	*Arid and Semi-Arid Region*	*Humid and Sub-Humid Region*	*Tropical Coastal and Island Region*
Main land-use systems	Grazing lands, forestry, horticultural and forestry species with agricultural crops, shifting cultivation in NE states	Agricultural crops (wheat, pulses, sugarcane, oil seeds), commercial wood lots, MPTS with crops and along boundaries degraded grazing lands	Trees and shrubs in crop fields and along Boundaries permanent grazing lands having trees and shrubs with stunt growth	Cash crops(tea and large cardamom), agricultural crops, shifting cultivation forests	Plantation crops and inter-cropping, field crops, fish culture, forestry.
Main land-use and ecological problems	Excessive defore-station, degraded lands, over-grazing, land sliding and shortage of fuel and fodder	Soil salinity, water logging, decline in soil fertility due to over-cultivation, fuel and fodder shortage	Drought, overgrazing, salinity, extension of sand dunes, low water tables, soil erosion, degraded lands	Deforestation soil erosion.soil acidity and consequent problems, decline in soil fertility, shortage of fuel and fodder. shortening of fallows	Deforestation and degradation of environment. coastal erosion, acid-soils and related soil problems, soil erosion along slopes. shortage of fodder
Major agro-forestry emphasis	Planting woodlots on hills, contour farming blended with MPTs. management of pastures, raising orchards	Soil reclamation using MPTs, NFTs, suitable grasses; growing NFTs/shrubs with crops along field boundary, raising woodlots on degraded lands	Raising shelter belts, wind breaks; growing NFTs in field, raising woodlots (NFTs) on degraded lands	Inter-cropping with nitrogen fixing trees, alley cropping, improved fallows	Afforestation of coastal regions with littoral and mangrove species, aquaculture with mangroves, multistoried cropping sytem; alley cropping

area and complicated legal procedures involved in tree farming trade and market fluctuations.

1. Poplar Based Agroforestry Practices

The farmers of Punjab, Uttar Pradesh, Haryana and parts of Uttarakhand have adopted Poplar based agroforestry in largescale and covering an area of 11.48 M ha (Chaturvedi and Khan, 2009). In North India it is a successful land use intervention for sustaining farm income to the farming community. Poplar can give yield of 150-200 m^3 per ha in block plantation and 12-20 m^3 per ha in boundary plantation at 8 years of rotation under scientific cultivation practices. In the inter spaces of Poplar plantation all the rabi and kharif crops are grown except rice.

2. *Eucalyptus* Based Agroforestry Practices

To meet the growing demand of population for pulpwood, plywood and timber industries, vast areas of Andra Pradesh, Haryana, Punjab, Uttar Pradesh, Uttarakhand, Odisha and many other parts of the country has been covered with Eucalyptus plantations. These plants are raised at a spacing of 3x2 m or 4x2 m. Many types of

3. Teak Based Agroforestry System

It is adopted for alley, bund and boundary or canal side plantation. Teak based Agroforestry models have developed by Agricultural Research Station, Dharwad involving arable crops (Sorghum, groundnut, chilli and ragi) silvicultural crop (teak), horticultural crop (papaya) and pastoral crops (subabul and grass). At end of 22-years, integration of teak and papaya with agricultural crops resulted in highest net returns (Rs.12,316 ha-1yr-1) followed by field crop+ teak+ papaya+ grass (Rs.10,977 ha-1yr-1) and field crops+teak+papaya+subabul (Rs.9,328 ha-1yr-1) as compared to field crop (Rs.3,546 ha-1yr-1).

4. Aonla Based Agroforestry System for Semi-Arid Region

Rainfed agriculture provides 44 per cent of food production from 68 per cent of total cultivable land and sustain 40 per cent of the population. In these regions except maize most of the coarse cereals are grown, which support 84 per cent groundnut, 91 per cent pulses, 55 per cent rice, 68 per cent cotton and 22 per cent wheat production. Productivity of these crops are one half to one third as compared to irrigated crops. Low productivity of seasonal crops promoted the evolution of sustainable land management system by integrating perennial woody components which are drought tolerant and gives early return with higher economic value. Aonla (*Emblica officinalis*) is the viable option among the semiarid and arid fruit crops. Fruit yielding starts after four years.

5. *Prosopis cineraria*-Based Agroforestry for Hot Arid Region of Gujarat and Rajasthan

To meet the demand of fodder and fuel wood for ever increasing human as well as livestock population more focus has been given in this diversified production system. Finally, tree-based production systems often produce crops of higher value than mono crops.

6. Subabul Based Agroforestry System

To meet the needs of paper and pulp wood requirements, this system is practiced in Andra Pradesh. Subabul plants are planted at a spacing of 1mx 1m, which give net return of Rs.1,68.632 ha^{-1} after three years of plantation. For fodder production spacing is 1.25m x 1.25m.

7. Multistoried Cropping System

In high rainfall and fertile areas of Keral multistoried cropping patterns integrating with perennial crops such as coconut, arecanut, rubber is intercropped with seasonal and annual crops of tapioca, banana, pulses and vegetables.

8. Agri-horticultural Systems

In North-Eastern region suitable Agri-horticultural systems are being adopted involving cultivation of ginger with fruit plants like mandarin and guava on gentle slopes. Pineapple also planted with multipurpose tree species in paired rows on hill slopes to check soil erosion and provides considerable cash income within a short period (Dhyani *et al.,* 1995).In this system in the inter spaces of fruit crops like lemon, sweet orange, mango, litchi *etc..* intercropping of cowpea, turmeric. blackgram *etc..* have also found to be promising (Dhyani, 2011).

Silvipastoral System

This system has been adopted to increase the productivity of degraded/ problematic soils to control soil erosion, conserve in-situ soil-moisture, yielding more fodder and fuel biomass. Silvipastoral system can be practiced on marginal lands with suitable fodder trees and grasses. A list of the fodder trees/shrubs/ grasses suitable for the specific area is presented in Table 12.5 (Singh and Chaturvedi, 2011).

Potential of Agroforestry for Sustainable Development

a) Productivity Enhancement

Agroforestry systems enhances the productivity of lands based on land capability classification. Biomass production of some multi-purpose tree (MPT) species is given in Table 12.6.

Potentials of silvipasture system in various regions of India under different agro-climatic regions of are based on research findings. Potential species and their production potential are given in Table 12.7 (Singh and Chaturvedi, 2011).

Table 12.5: Promising Trees, Grass and Legumes for Silvipastoral Systems in Few Regions of India

Agroclimatic Zones	*Fodder Trees*	*Grass and Legumes*	*References*
Temperature, (J&K, Himachal Pradesh and NW Himalaya)	**West regions** *Rabinia pseudoacacia, Geltus austratis, Papulus alba, Quercus semecarpifolia, Q.dilalata, Acer campbelli*	*Lucerne (Medicago satira), Red clover (Trifolium pretense), White clover (T. repens), Cocksfoot (Dactylis glomerala), Timothy(Pheleum pratense)*	Singh (1986)
	Dry region *Salix fragilis, P. ciliate, P. euphratica, Belula utilis, Acer candatum, Q.ilex*		
Kangra Valley, HP	*Erythrina suberosa Dalbergia sissoo, A.catechu, Albizia stipulata*	*Eulaliopsis binata, Chrysopogonfulvus, Bauhinia purpurea, Albizia lebbek, Napier bajra, Setaria anceps*	Vishwantham *et al.* (1998)
North-eastern Hill region	i) *Albizia lebbek, Bauhinia purpurea, leucaena leucocephala.* ii) *Salix spp., Populus ciliate, Castanopsis, Celtis australis, Sachima/vallichii, Ficus spp., Bauhinia variegate*	*Panicum maximu, Setaria sphacealata, Pennicetum polystachyon, Stylosanthes guyanensis, Bajra x Hybrid Napier, Panicum maximum, Doctylis glorneruta, Lolium perenne*	Singh *et al.* (1996)
High rainfall, Doon Valley	*Leucaecaena leucocephala, Grewia optiva, Acacia sp., Bauhinia retusa, Salix tetrasperma*	*Eulaliopsis binata, Chrysopogon fulvus*	Dhyani *et al.* (1988)
Saline-alkali soils	*Prosopis juliflora, Acacia nilotica, Terminalia qrjuna, Pongamia pinnata*	*Leptochloa fusca, Chloris gayane, Brachiana mutica, Sporobolus spp.*	Singh *et al.* (1994)

Table 12.6: Biomass Production of different Tree Species

Tree Species	*Productivity (t ha^{-1})*
Casuarina equisetifolia	171
Leucaena leucocephala	159
Eucalyptus hybrid	152
Dalbergia sissoo	65
Albizia lebbeck	79
Albizia procera	100
Madhuca latifolia	41
Hardwickia binata	42

Table 12.7: Potential of Silvipasture System in Various Regions of India

Region	*Components*	*Results*
Alpine Region	Only pastures with suitable grasses and legumes	*Agrostis spp., Chrysopogon gryllus, Dactylis glomerata, Poa pratensisTrifolium repens, T.pratense, Medicago falcata*
Temperate Region	Intensive pasture production, temperate horticultural plant combined with pasture legumes as ground floor, fodder forest legumes	5.7 tha^{-1}, on an av. 8 to 10 tree^{-1} dry matter
North-eastern Region	Silvipasture, i) *Alnus nepalensis+ Stylosanthes guianensis + Panicum maximum* *ii) Ficus hookerii + Thysanolaena maxima*	 13.5 tha^{-1}(DM) fodder yr. 12.3 ha^{-1} fodder yield
Outer Himalayas Region	Silvipasture systems, *Dalbergia sissoo+Chrysopogon fulvus* *Acacia catechu + Eulaliopsis binata* *Leucaena leucocephala (high density)* + Hybrid Napier (NB 5) *Bauhinia purpurea, Albizia lebbeck + Eulaliopsis binata*	Suitable system for utilization of degraded lands for fuel, fodder and fibre 64-71 tha^{-1} fuelwood5.2-5.5 tha^{-1} yr^{-1} fodder grass for short rotation of 4-5 years Most suitable for gravelly riverbed lands of Doon valley
Shiwalik foothills	Silvipasture, Fodder production from wastelands and watershed management	Fodder production increased by 267 per cent

Region	*Components*	*Results*
Indo-Gangetic plains	Tree-grass combination	Annual production of 8-10 tha^{-1} (4.5-5t ha^{-1} fuelwood) 7-10t $ha^{-1}yr^{-1}$ biomass production
	Forage production on dry lands	Leucaena + sorghum - safflower-Forage yield of 12.3 t ha^{-1}
Humid and sub- humid	Use of quick growing leguminous fodder plants in crop rotations, plantation crops with pastures, farm forestry.	Best for pasture development in the eroded marginal and sub- marginal hilly area of Southern Bihar
	Deenanath (*Pennisetum pedicellautm*) +*Stylosanthes spp.* and Deenanath + *Leucaena spp.*	
	i) *Leucaena spp.* (Hawaiian, Hybrid-28) + fodder grass. *Glyricidia/Sesbania/*	Suitable combination for plain area in West Bengal
	ii) *Ficus hookerii*	Suitable for hilly areas
	Litsea+Thysanolaena/Andropogon nardus	
Coastal area (kerala *etc.*.)	Fodder species(Subabul/Agasti/ Desmanthus) + Congo signal + *Stylosanthes or Centrosema.*	Total fodder yield of 22-29 t ha-1 could be obtained in coconut Based fodder production systems.
Arid and semi- arid region	Grazing land and silvipasture and Hortipasture in arid and integration of agri-silvipasture,	Very good potential for silvipastoral enterprise
	Forage-cum-cropping farming on the marginal and sub-marginal lands with intercropping of dryland cereals and legumes, *Acacia tortilis, Albizia lebbeck, Prosopis cineraria* with *Chrysopogon fulvus, Cenchrus ciliaris, C. setigerus.*	

b) Protection and Conservation of Natural Resources

Agroforestry practices reduce erosivity of rainfall/runoff and erodibility of soil through dissipation of rainfall energy by canopy at low height, surface litter, obstructing overland flow, root binding and improving soil physico-chemical conditions (Dhyani, 2011). A summary of agroforestry practices with potential for the control of soil erosion is given in Table 12.8 (Young, 1997).

Table 12.8: Agroforestry Practices with Potential for Control of Soil Erosion

Agroforestry Practices	*Suitable Climates*	*Important Points*
Multistory including home garden	Mainly in Humid and moist sub-humid climates	Possess an inherent capacity to control erosion through combination of herbaceous cover with abundant litter
Plantation crop combinations	Humid to moist sub-humid climates	Densely planted combination of agricultural plantation crops with MPTs appear to control erosion effectively on at least moderate slopes
Hedgerow intercropping, barrier hedge	Humid, sub-humid, possibly in semi-arid climates	A considerable apparent potential to combine erosion control with arable use on gentle to moderate slopes,more speculative potential on steep slopes.
Trees and erosion control structure	All climatic conditions	Supplementary use of trees stabilizes earth structures and gived production from land they occupy.
Windbreaks and shelterbelts	Semi arid zone	Proven potential to reduce wind erosion.
Silvipastoral practices	Semi arid, humid and sub-humid	Opportunities for inclusion of trees and shrubs as part of overall programmes of pastoral improvement
Reclamation forestry	All climatic conditions	Potential for planted design and development.
Combinations of the above in integrated watershed management	All climatic conditions	Substantial opportunities to include agroforestry with other major kinds of land use in integrated planning and managements

c) Maintenance of Soil Health

Trees in agroforestry helps in maintaining soil fertility by adding organic matter and checking soil erosion. Ameliorative effect of agroforestry has also been reported for problematic soils such as salinity, alkalinity and waterlogging problems (bio-drainage) are presented in Table 12.9.

Table 12.9: Effective Trees for Ameliorating Soil Health

Alkali Soils	*Saline Soils*	*Bio-drainage*
Prosopis juliflora	*Casuarinas equisetifolia*	*Populous deltoids*
Acacia nilotica	*Leucaena leucocephala*	*Terminalia arjuna, Eucalyptus tereticornis, Acacia auriculiformis*
Tamarix articulate	*Prosopis juliflora*	*Syzigium cuminii*

Alkali Soils	*Saline Soils*	*Bio-drainage*
Eucalyptus tereticornis	*Acacia nilotica*	*Albizia lebbeck*
Pongamia pinnata	*Tamarix troubii*	*Pongamia pinnata*
Terminalia arjuna	*Tamarix articulate*	*Dalbergia sissoo*
Dalbergia sissoo	*Acacia farnesiana*	
	E. Camaldulensis	

d) Amelioration of Microclimate

Microclimate under the canopies of trees in agroforestry reduces soil and air temperature, irradiance and wind velocity. These changes directly influence the crop growth and productivity.

e) Carbon Sequestration

Agroforestry can store carbon by increasing the production of soil organic matter and biomass production which can pool more CO_2 from the air. In India, majority of population depends on agriculture. Agroforestry provides a long term strategy to increase the amount of carbon while still allowing for growing of food crops. Average sequestration potential of agroforestry has been estimated to be 25 t C ha^{-1} over 96 M ha of land in India (Sathaye and Ravindranath, 1998). Soil carbon storage and mitigation potential in different land use systems is presented in Table 12.10.

Table 12.10: Soil Organic Carbon (SOC) Content in some Agroforestry Systems

Region	*Agroforestry System and Components*	*Total Storage (t C ha^{-1})*	*References*
Semi-arid	Silvipastoral (age 5 yrs)		Rai *et al.* (2001)
	A. nilotica + natural pasture	9.5-17.0	
	A. nilotica + established pasture	19.7	
	D. sissoo + natural pasture	12.4	
	D. sissoo + established pasture	17.2	
	H. binata + natural pasture	16.2	
	H. binata + established pasture	17	
North-western India	Silvipastoral (age 6 yrs)		Kaur *et al.* (2002)
	Acacia/Dalbergia/Prosopis +Desmostacya	6.8-18.55	
	Acacia/Dalbergia/Prosopis +Sporobolus	1.5-12.32	
Central India	Block plantation(age 6 yrs)		Swamy *et al.*, 2003)
	Gmelina arborea	24.12-31.12	
Arid region (Rajasthan)	Agri-silvicultural (age 8 yrs)	12.7-13.0	Singh (2005)
	E.officinalis + V.radiata		

Region	*Agroforestry System and Components*	*Total Storage ($t\ C\ ha^{-1}$)*	*References*
Semi-arid region	Agri-silvicultural (age 11 yrs)	26.02	Anon. (2005)
	D. sissoo + crop		
Semi-arid region	Agri-silvicultural (age 5 yrs)		Ram Newaj *et al.* (2006)
	A. procera + black-gram-mustard	34.77	
	A. procera + green-gram-wheat	35.13	
North-western Himalaya	Silvipastoral system	2.17	Anon. (2006)
	Agri-hortipastoral	1.15	
	Hortipastoral	1.08	

f) Employment Generation and Livelihood Security

Agoforestry can enhance the income of farmers by providing more employment opportunities for livelihood security. A large number of profitable combinations of tree-crop are available in different agro- climatic zone. The B:C ratio of some of the important agroforestry system is given in Table 12.11.

Table 12.11: Agroforestry Systems with Respective B:C Ratios in different Agroclimatic Zones

Zone	*Tree*	*Crops*	*Yrs*	*B:C Ratio*
E.Himalaya	*Anthocephalus kadampa*	Paddy	8	1.6
W.Himalaya	*Prunus persica*	Maize, Soya bean	15	1.21
Lower Gangetic	*Eucalyptus hybrid*	Paddy,wheat	10	2.33
Middle Gangetic	*Dalbergia sissoo*	Sesamum	20	1.15
Upper Gangetic	*Populous deltodes*	Paddy,wheat	7	3.31
Trans Gangetic	*Acacia nilotica*	Pearl Millet	20	2.61
Eastern plateau	*Harwickia binate*	Pennisetum	15	2.13
Central plateau	*Emblica officinalis*	Ground nut	8	1.01
Western plateau	*Albbizia amara*	Cenchrus	16	1.56
Southern plateau	*Tamarindus indica*	Chilli	10	2.03
West Coastal Ghat	*Casuarina equisetifolia*	Paddy	7	1.39
Gujarat plains	*Dalbergia sissoo*	Caster	20	2.14
Western dry	*Prosopis cineraria*	Pearl millet	20	1.65
Island	*Cocos nucifera*	Paddy	6	1.17

Conclusion

Agroforestry generates more integrated, diversified, sustainable land use systems and only viable option to increase the country's forest and tree cover to the 33 per cent. Ecosystem services as well as monetary gain from different agroforestry systems have already been established and need to be focused for attracting farmers to adopt agroforestry along with the improved technology for sustainability. In this context, there is an urgent need for large-scale availability of improved and qualitative planting stocks of suitable agroforestry species for rapid expansion of area under agroforestry systems. In Indian region, Agroforestry has extensive scope and significant opportunity that can only fit to resolve the problems of food availability and climate change.

REFERENCES

Anonymous (2006) All India Coordinated Research Project on Agroforestry. NRCAF, Jhansi.

Bijalwan A, Sharma CM and Kediyal VK (2011) Socioeconomic status and livelihood support through traditional agroforestry systems in hill and mountain agro-ecosystem of Garhwal Himalaya, India. The Indian Forester 138(12): 1423-1430.

Chaturvedi OP and Khan IA (2009) Agroforestry research perspectives and potential. In: Agroforestry: natural resource sustainability, livelihood and climate moderation (Eds. Chaturvedi O.P, Venkatesh A, Dwivedi R.P, Singh R and Dhyani S.K. Satish Serial Publishing House, Delhi, India, pp.27-42.

Chauhan MK, Ram J and Dagar JC (2012) Biodrainage and Carbon Sequestration. Lambert Academic Publishing, pp.304

Chavan SB, Keerthika A, Dhyani SK, Handa AK, Newaj R and Rajarajan K (2015) National agroforestry policy in India: a low hanging fruit. Current Science 108(10): 1826-1834.

Cittapur BM and Patil DK (2017) Ecosystem services rendered by tree-based land use systems. Indian Journal of Agricultural Sciences 87 (11): 1419-1429.

Dagar JC (1995) Agroforestry systems for the Andaman and Nicobar Islands. Int. Tree Crops J 8: 107–128.

Dagar JC, Singh AK and Arunachalam A (2014) Introduction. In: Agroforestry Systems in India: Livelihood Security and Ecosystem services (eds.). Advances in Agronomy 10: 1-20.

Dagar JC (1991) Agroforestry systems in humid and sub-humid regions of India. In : Proc. Sym. Farming Systems for Sustained Productivity in Humid Tropics. CARI, Port Blair. pp.51-63.

Dagar JC (2000) Biodiversity of saline habitats. In: Environment, Biodiversity and Conservation (Eds. MA Khan and S. Farooq). APH Publishing Corp., New Delhi, pp. 135-162.

Dhillon WS, Srinidhi HV, Chauhan SK, Singh C, Singh N and Jabeen N (2010) Microenvironment and physiology of turmeric cultivated under poplar tree canopy. Indian Journal of Agroforestry 12(2): 23-37.

Dhyani SK (2011) Alternate land use and agroforestry systems for resource conservation and enhanced productivity in hills. Proceedings of workshop on mountain agriculture in Himalayan region: Status, constrains and potentials, pp. 107-132.

Dhyani SK, Chauhan DS, Singh BP and Prasad RN (1995) Tree effect on yield and quality of pineapple(Kew) in a silvi-horti systems of agroforestry in NEH region of India. Range management and Agroforestry 16(2): 89-93.

Dhyani SK, Dadhwal KS and Katiyar VS (1988) Biological measures for rehabilitation of abandoned mine area. Indian Farming 38(7): 15-18.16

Garbyal SS (1999) Jhuming (Shifting cultivation) in Mizoram (India) and new land use policy- how far it has succeeded in containing this primitive agriculture practice. Indian Forester 125: 137-148.

Gill RIS, Singh B, Kaur N and Sangha KS (2010) Agroforestry - A viable option for crop diversification in Punjab. Department of Forests and Wildlife Preservation, Punjab, pp. 20.

Hegde NG (2005) Tree based Farming for Sustainable Livelihood at the workshop on Tropical Forestry and Poverty Eradication, Bangkok, Thailand.

Jamala GY, Shehu HE, Yidau JJ and Joel L (2013) Factors influencing adoption of agroforestry among small holder farmers in Toungo, Southeastern, Adamawa State, Nigeria. Journal of Environmental Science, Toxicology and Food Technology 6: 66-72.

Kareemulla K, Rizvi RH, Yadav RS, Munnaram and Dhyani SK (2009) Agroforestry for rural development – cooperatives as a viable institution. Agricultural situation in India, pp. 61-67.

Kaur B, Gupta SR and Singh G (2002) Carbon storage and nitrogen cycling in silvipastoral system on a sodic soil northwestern India. Agroforestry Systems 54: 21-29.

Kumar AK (2010) Carbon sequestration: underexplored environmental benefits of Tarai agroforestry. Indian Journal of Soil Conservation 38: 125-131.

Lundgren and Raintree (1982) Sustained Agroforestry. In: Agricultural Research and Development-Potential and Challenges in Asia, bnestel (Ed), ISNAR, The Hague, pp.37-49.

Mathukia RK, Sagarka BK and Panara DM (2016) Fodder production through agroforestry: a boon for profitable dairy farming. Innovare Journal of Agri. Sci 4(2): 13-19.

Mbow C, Noordwijk MV, Luedeling E, Neufeldt H, Minang PA and Kowero G (2014) Agroforestry solutions to address food security and climate change challenges in Africa. Current Opinion in Environmental Sustainability 6: 61–67.

National Agroforestry Policy. Department of Agriculture and Cooperation, Ministry of Agriculture, Government of India, 2014.

Nerlich K, Graeff-Honninger S and Claupein W (2013) Agroforestry in Europe: A review of the disappearance of traditional systems and development of modern agroforestry practices, with emphasis on experiences in Germany. Agroforestry Systems 87 (2): 475-492.

Pandey DN (2007) Multifunctional agroforestry in India. Dogra, A.S. (2007). Contribution of trees outside forests toward wood production and environmental amelioration. Indian Journal of Ecology 38: 1-5.

Pathak PS, Slanki KR, Rai P, Handa AK and Pateria HM (2006) Agroforestry. In: Handbook of Agriculture. Indian Council of Agriculture Research, N. Delhi.

Pathak PS and Dagar JC (2000) Traditional agroforestry systems: an overview. In: Soni P, Srivastava AK (eds) Landmarks of botany in India. Suriya International Pub., Dehradun, pp 26–49.

Rai SN and Chakrabarti SK (2001) Demand and supply of fuelwood and timber in India. Indian Forester 127: 263-279.

Newaj R, Shanker AK and Yadav RS (2006) Carbon and Nitrogen dynamics in *Albizia procera* based agroforestry system. Annual report of ICAR ad-hoc scheme, NRCAF, Jhansi

Rowbotham CJ (1924) The taungya system in Cachar Division. Ind. For. 50: 356-358.

Sathaye JA and Ravindranath NH (1998) Climate change mitigation in the energy and forestry sectors of developing countries. Annu. Rev. energy Envirn. 23: 387-437.

Singh BP, Dhyani SK and Prasad RN (1994) Traditional agroforestry systems and their soil productivity on degraded alfisols/ultisols in hilly terrain. In: Agroforestry systems for Degraded Lands (Eds. P. Singh, P.S.Pathak and M.M. Roy). Oxford and IBH Pub. Co. Pvt. Ltd., New Delhi, pp. 205-214.

Singh KA, Yadav BPS and Goswani SN (1996) Forage resource management for sustained livestock productivity in Himalayan Agro-ecosystems. Research Bulletin No. 39, ICAR Research Complex for NEH Region, Barapani, pp.94.

Singh P (1986) Forage production in India, RMSI, IGFRI, Jhansi, pp.213

Singh RP (1997) Alternate land use systems for sustainable development. In Silvipastoral systems in arid and semi -arid ecosystems (Eds Yadav M.S, Singh M., Sharma S.K, Tewari J.C and Burman U) CAZRI, Jodhpur, India, pp. 249-266.

Singh RP and Osman M (1995) Alternate land use systems for drylands. In Sustainable development of dry land Agriculture in India (Ed Singh R.P) Scientific Publishers, Jodhpur, pp. 375-398.

Singh C and Chaturvedi OP (2011) Rehabilitation of degraded lands through silvipastoral systems-a balanced approach for alternative land use in the north-Western Himalayas. In: Agroforestry systems for resource conservation and livelihood security in Lower Himalayas (eds. Panwar, P, Tiwari, A.K. and Dadhwal, K.S.) New India Publishing Agency, New Delhi, pp.273-296.

Singh KA (2008) Resource management perspective for forage production and agroforestry system development in eastern Himalayan region: A review. Indian journal of Agronomy 53(4): 255-266.

Singh G (2005) Carbon sequestration under an agrisilvicultural system in the arid region. Indian Forester 131 (4): 543-552.

Swamy SL, Puri SS and Singh AK (2003) Growth biomass, carbon storage and nutrient distribution in *Gmelina arborea* Roxb. Stands on red lateritic soils in central India. Bio resource technology 90: 109-126.

Thenmozhi S and Thilagavath P (2014) Impact of agriculture on Indian economy. International Research Journal of Agriculture and Rural Development 3.

Tiwari SK (2008) Agroforestry Project; Dept. of Genetics and Plant Breeding College of Agriculture, G.B. Pant University of Agriculture and Technology Pantnagar - 263145.

Verchot LV, Van-Noordwijk M, Kandji S, Tomich T, Ong C, Albrecht A, Mackensen J, Bantilan C and Palm C (2007) Climate change: Linking adaptation and mitigation through agroforestry. Mitigation and adaptation Strategies for Global Change 12: 902: 918.

Vishwantham MK, Samra JS and Sharma AR (1998) Biomass production of silvipasture systems on gravelly lands of Doon valley. Bulletin No. T-39/D-26, CSWCRTI, Dehradun, pp.126.

World Bank. 2007. Jharkhand - Addressing the Challenges of Inclusive Development. Report No. 36437-IN, pp.148.

Young A (1997) Agroforestry for soil management. 2nd Ed. C.A.B. International, Wallingford, UK, pp.320.

Chapter 13

Natural Product Mediated Therapeutic Application of Artificial Membrane Models: An Insight

Gayatree Panda and Santosh Kumar Sahu

Department of Biotechnology, Maharaja Sriram Chandra Bhanja Deo University, Baripada, Mayurbhanj–757003, Odisha
e-mail: drsantoshnou@gmail.com

Introduction

Artificial membrane models or liposomes are membrane-bound entities with one of more lipid bilayers that enclose single or multiple aqueous compartments. Liposomes are used to deliver both hydrophilic and lipophilic chemotherapeutics against multiple human disorders such as cancer and pathogen infection. In addition, liposomes are ideal vehicles for the delivery of vaccines and enzymes. Based on the number of bilayers present and methods, liposomes can be unilamellar, vesicles, multilamellar vesicles, and giant unilamellar vesicles. Here, the term "vesicle" refers to a specific kind of compartment that is encircled by an aqueous medium (from the Latin vesicula, -ae, little bladder). A vesicle's boundary is made up of one (or more) thin layers or membranes made of amphiphilic molecules, and its interior is a tiny watery entity. In a vesicle membrane, the amphiphiles are decorated so that the hydrophilic components are exposed to the aqueous medium and the hydrophobic components have accumulated to form the interior of each layer. Lipids such as cholesterol, phosphatidylcholine and bacterial phospholipids are amphipathic molecules with water-loving and water-hating parts. However, natural phospholipids, modified natural phospholipids, semisynthetic phospholipids, synthetic phospholipids, phospholipids with non-natural head groups were utilized. Liposomes are composed of one or more lipid bilayers created through hydrophilic and hydrophobic interactions with the surrounding aqueous environment. In this context, the hydrophobic regions (tails) of liposomes are repelled by water molecules, leading to the self-assembly of liposomes (Valenzuela SM., 2007). In liposome preparation, the types and amounts of phospholipid, the ionic and charge properties of the aqueous medium, and the hydration period are critical factors that influence the final liposome structure.

There are several methods for preparation of these structures including electroformation using copper electrode, ultrasonication, extrusion through polycarbonate filters, freeze-thawing, ethanol injection, detergent removal method *etc.*. Based on lamellarity, unilamellar and multilamellar vesicles were formed in which multilamellar vesicles are artifacts, thus avoided. Unilamellar liposomes are spherical vesicles made of a single lipid bilayer that serve as an effective model for natural membranes and can be utilised to study membrane constituents (Bangham and Horne, 1964). This article details classification, formation, and benefits of liposomes, including improved synthetic drugs as well as natural phytodrug solubility, sustained release, site-specific targeting, and enhanced tissue penetration facilitating the novel drug delivery to the target sites. The Similpal Biosphere Reserve (SBR), an enriched biodiversity situated in Northern regions of Odisha serves as a huge resource of pharmaceutical natural products with several anti-diabetic, anti-bacterial, anti-cancer, anti-malarial, anti-rheumatic, anti-asthma, anti-fungal effects (Panda *et al.,* 2017). These natural drugs can be trapped within the liposomes and administered in various ways to treat several diseases with minimal or no side effects.

Major Types of Liposomes and Factors Affecting Liposome Formulation.

The desired vesicles when using liposomes for drug delivery typically range in size from 50 nm to 150 nm.

1. Small Unilamellar Vesicles (SUVs) and Large Unilamellar Vesicles (LUVs)

The unilamellar vesicles such as SUVs (Small Unilamellar Vesicles) of size 20-100nm and LUVs (Large Unilamellar Vesicles) of size >100nm are composed of conventional biomembrane phospholipids, which have one hydrophilic head group and two lipophilic chains (Mulla *et al.,* 2019). The vesicle membrane is built from one single bilayer with the head groups of the lipids of the inner layer facing towards the interior of the vesicles and the head group of the outer layer being in contact with the external aqueous medium. This structure facilitates in an even distribution of trapped agents within a single internal aqueous compartment.

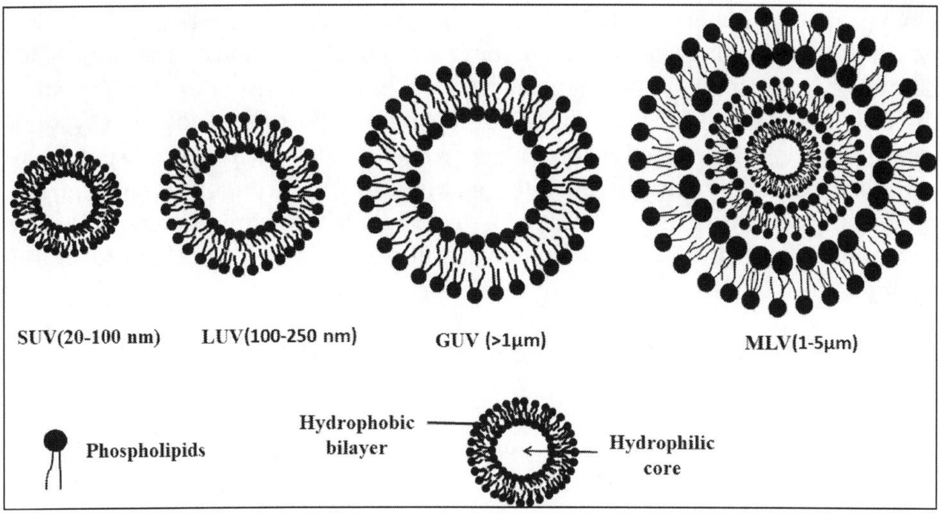

Figure 13.1: Schematic Representation of different Types of Artificial Membrane Models or Liposomes According to their Size and Lamellarity.

2. Giant Unilamellar Vesicles (GUVs)

Giant Unilamellar Vesicles (GUVs) are unilamellar vesicles with a diameter exceeding 1μm, composed of polar lipids. Their dimensions are analogous to those of eukaryotic cells, hence emulating the identical lipid reservoir and membrane curvature (Fenz and Sengupta, 2012). Furthermore, their substantial size, stability, meticulous modulation of lipid composition, integration of proteins (Jorgensen *et al.*, 2016), and the internal development of an underlying cytoskeleton (Tsai *et al.*, 2011) facilitate the observation of cellular components using optical microscopy. GUVs were visualised via phase contrast microscopy.

In a unilamellar structure, liposomes possess a monophospholipid bilayer, whereas in a multilamellar structure, they exhibit an onion-like configuration. Multilamellar vesicles consist of many concentric phospholipid spheres formed by the creation of numerous unilamellar vesicles within larger liposomes. For hydrophilic compounds alone, the encapsulation effectiveness of liposomes increases with liposome size and decreases with the number of bilayers (Ong *et al.*, 2016). The circulatory half-life of liposomes is markedly affected by the vesicle size. The quantity of the entrapped medicine is affected by the size and number of bilayers. Numerous theories elucidate the interaction between liposomes and cell membranes, encompassing targeted or nonspecific endocytosis (altered by receptor-mediated endocytosis), local fusion (adhesion), phagocytosis, and absorption into the cell membrane (Kelly *et al.*, 2011; Caracciolo *et al.*, 2005; Storm *et al.*, 1995). Multiple factors influence the interaction of liposomes with cells, including composition (Duzgunes *et al.*, 1999), liposome diameter, surface charge (Lee *et al.*, 1993), targeting ligands on the liposome surface, and the biological environment (Kono *et al.,* 1999). Prior to vesicle generation, a critical physicochemical component to consider is the phase transition temperature. This parameter is intimately associated with membrane fluidity and stability, since the transition temperature of phospholipids influences several colloidal systems (Matsuki *et al.,* 2017). Additional parameters, such as surface charge (zeta potential) and particle size, might influence the stability of the formulation and indicate the orientation of molecules within the vesicle, including amino acids (Disalvo and Bouchet, 2014; Morin *et al.,* 2015). A decrease in zeta potential results in aggregation, facilitating a stable liposome formulation.

Different Preparation Methods of Liposomes

1. Natural Swelling

This approach involves the growth of vesicles from stacked lipid bilayers, driven by osmotic pressure, electrostatic interactions, and the hydrophobic effect. In this process, lipids dissolved in chloroform are applied to a solid substrate, and subsequent to solvent evaporation, several stacks of bilayers are formed by clustering. Following the addition of the buffer solution, vesicles may be acquired after several days. Furthermore, an augmented distance between the lipid layers (Akashi *et al.,* 1996), resulting from temperature fluctuations and electrostatic repulsion of the lipid head-groups, facilitates a more efficient inflow of the buffer solution between the bilayer stacks. Consequently, incorporating a certain quantity of charged lipids and sucrose between the lipid layers to enhance osmotically driven flow can augment the GUV production (Tsumoto *et al.,* 2009). However, the disadvantage is that a substantial quantity of multilamellar vesicles and lipid debris is generated, resulting in a reduced yield of GUVs. These are undesirable giant unilamellar vesicles featuring interior tubules and affixed tiny vesicles.

2. Electroformation

This technique yields substantial quantities of vesicles after a few hours in distilled water or sucrose. Lipids are deposited on two electrodes, and the swelling of the dried lipid film into vesicles is augmented by an electric field due to the interplay of electrostatic interactions, redistribution of bilayer counter ions, membrane surface dynamics, and electro-osmotic effects (Dimitrov and Angelova, 1987). When an electric field is applied, the lipids respond to the alternating electric current based on their net charge. The non-covalently associated counter ions of the lipids redistribute between the bilayers in response to the electric field. The inflow of water between the bilayer stacks is augmented, causing the split membrane to protrude.

3. Microfluidics

The buffer solution is injected onto a pre-assembled lipid bilayer at a water-oil interface using a micronozzle or micropipette, resulting in the swelling of GUVs from this bilayer. This technique was enhanced by Stachowiak *et al.* (2009) by the use of a piezoelectric actuator from an inkjet printer. The double emulsion approach (Shum *et al.,* 2008; Shah *et al.,* 2008) involves the formation of water-in-oil-in-water droplets, which then undergo solvent evaporation, resulting in the creation of GUVs from the double emulsion droplets. Protein expression was achieved via vesicles (Teh *et al.,* 2011). Nonetheless, nonphysiological polymer solutions may introduce artefacts in membrane diffusion investigations.

4. Sonication

This method involves placing MLV into a bath sonicator or disordering them with the use of a probe sonicator. The application of ultrasonic irradiation reduces vesicle size by providing energy for the lipid suspension of MLV.

5. Freeze-Drying

The most adapted mechanical dispersion technique, known as "lipid-film hydration," creates a thin film made of a phospholipid and cholesterol membrane by evaporating the organic solvent from a phospholipid and cholesterol solution. With the help of vortexing, sonication, and hydrating the evaporated solution with a phosphate buffer solution, liposomes are produced (Liu *et al.,* 1994). Phytochemical-loaded liposomes named phytosomes are prepared using this method. Briefly, phytochemicals or bioactive compounds are typically combined with phospholipids in an appropriate non-polar organic solvent.

The solution is dried and concentrated under reduced pressure through rotary evaporation to create a thin film. The film is subsequently hydrated. The thin film comprises phytosomes that precipitate upon the addition of an aliphatic hydrocarbon, such as n-hexane. Phytosomes are subsequently collected by solvent removal, via lyophilization or spray drying (Dutt *et al.,* 2023).

6. Reverse Phase Evaporation

The lipids are immersed in an organic solvent, which is subsequently combined with an aqueous buffer and the hydrophilic medication. The organic solvent then evaporated using a rotary evaporator, resulting in the formation of lipid vesicles inside the aqueous solution. Extrusion can diminish the average size and dispersibility of the synthesised vesicles (Himanshu *et al.,* 2011; Bhupendra *et al.,* 2015). A continuous wave laser at 532 nm is employed for illumination in fluorescence imaging of these artificial membranes. The beam is magnified fourfold with a telescope and directed onto the rear focal plane of an oil-immersion objective (60x or 30x, Olympus), producing wide-field illumination on the specimen. The fluorescence light is gathered by the objective and reflected by a beam. A variety of fluorescent dyes, including NBD-PE, Rhod-PE, SRB, and Carboxyfluorescein, were employed to assess fluorescence scattering.

Table 13.1: Advantages and Drawbacks of Several Methods Employed in Liposome Formulations and Utility

Types of Liposome Formation	*Advantages*	*Disadvantages*	*Reference*
Natural swelling	A gap between the lipid layers may allow the buffer solution to enter the bilayer stacks more effectively.	Multilamellar vesicles and lipid debris are formed along with fewer GUVs.	Tsumoto *et al.,* 2009
Electroformation	Higher GUV yield due to the even and periodic redistribution of electric fields.	Agarose sealing could be compromised, revealing an agarose-containing swelling vesicle inside the electroformation chamber.	Lira *et al.,* 2014
Microfluidic channel method	Microdomain formation is facilitated by ultrathin, double-emulsion vesicles with little residual solvents.	A substantial amount of lipid is necessary. Residual solvents are confined inside the bilayer, making the GUVs inappropriate for diffusion investigations or domain formation.	van Swaay and deMello, 2013
Sonication	Temperature of the dispersed lipid is easily controllable.	Low internal volume/ encapsulation efficacy, easy rupture of phospholipids and chemicals that have been encapsulated, removal of big molecules, metal contamination from the probe tip and the presence of MLV in addition to SUV are some of the problems.	Kataria *et al.,* 2011
Extrusions	Resulting liposomes are rather larger than sonicated SUVs.	High temperature is difficult to obtain and the working volumes are comparatively small.	Himanshu *et al.,* 2011
Freeze Drying	Encapsulation efficacies ranging from 20 per cent to 30 per cent were achieved.	Sugars, high ionic strength, and divalent cations interrupt the freezing.	Akbarzadeh *et al.,* 2013
Reverse phase evaporation	Advantageous for high molecular weight compounds.	Therapeutic peptides may be denatured due to organic solvents and sonication conditions.	Szoka *et al.,* 1980

Clinical Application of Liposomes

1. Encapsulation of Drugs by Liposomes and their Therapeutic Benefits in Various Fields

Encapsulation of hydrophilic pharmaceuticals yields a combination of hydrated lipids and hydrophilic compounds. Drugs can penetrate the liposome core while other substances stay external to the liposome. Residual materials will eliminate drug entrapment in liposomes. Gel filtration column chromatography and dialysis are employed to purify these two components (drugs and residual external contaminants). The phospholipid bilayer of liposomes serves as a domain for hydrophobic drug encapsulation. The trapping of antibiotics like verteporfin will diminish their migration towards the outside aqueous and inner regions of liposomes. These pharmaceuticals are encapsulated via the solubilisation of the antimicrobials in an organic solvent and phospholipids.

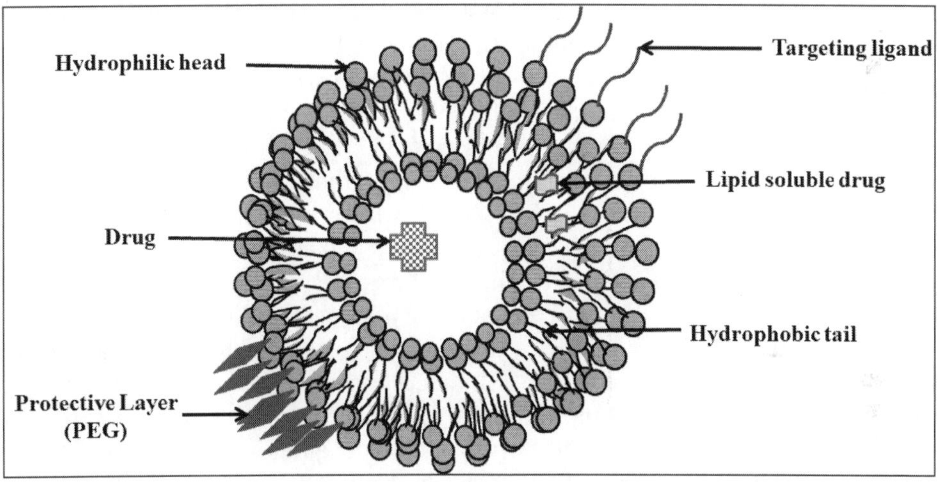

Figure 13.2: Schematic Diagram Illustrating an Encapsulation of both Hydrophilic and Hydrophobic Drugs in the Aqueous Core and Lipid Bilayer within the Liposome, Respectively.

The surface of the liposome allows for the attachment of targeting ligands and a polyethylene glycol (PEG) coating for active and passive targeting, respectively.

Subsequently, laser light can be employed for drug activation. Significantly, recent developments in this domain have often employed giant vesicles characterised by little complexity regarding lipid composition and buffer conditions to examine membrane shrinkage or invagination processes (Furusato *et al.,* 2018) and shape regulation (Fanalista *et al.,* 2014). Other well documented membrane biological activities using giant vesicles include the transport and release of molecules across membranes (Parui *et al.,* 2019) and the interaction of proteins and peptides with membranes (Yang *et al.,* 2007). The interaction of antimicrobial peptides with lipid vesicles that simulate bacterial membranes has demonstrated a lipid-dependent and ion-dependent shift in membranes. The lipid vesicles contain biologically inert phospholipids having immunogenicity and low inherent toxicity.

As nanocarriers of anticancer drugs, indium 111-labelled liposomes could accumulate within solid tumours such as malignant lymphoma and Kaposi's sarcoma (Allen *et al.,* 2013). As measured by flow-assisted cell sorting in pancreatic adenocarcinoma (PANC-1) and insulinoma (INS-1) cells, the targeting capacity of liposome-protein complexes strongly correlates with cellular absorption (FACS). The outcomes demonstrated that cellular uptake of the liposomal formulation with the greatest abundance of protein corona fingerprints (PrCFs) was significantly greater than that of Onivyde, an irinotecan liposomal medication approved by the Food and Drug Administration in 2015 for the treatment of metastatic pancreatic ductal adenocarcinoma (PDAC) (Palchetti *et al.,* 2019). Furthermore, using liposomes to gather the PrC coating layer from human plasma proteins allowed Digiacomo *et al.,* to discover a possible protein biomarker for PDAC (Digiacomo *et al.,* 2021). Cardiotoxicity is one of the undesired side effects associated with doxorubicin treatment; that was clinically reduced by doxorubicin encapsulated liposome as the entrapped doxorubicin cannot become directlyaccumulate at the myocardium and cardiomyocytes (Bulbake *et al.,* 2017). Additionally, Doxorubicin encapsulated liposome forms a nonflexible bilayerthat enhanced drug retention, thereby, gain approval from the FDA in 1995 for the treatment of Kaposi's sarcoma in patients suffering from AIDS and recurrent ovarian cancer (Bozzuto and Molinari., 2015). The liposomes with the least toxicity contained neutrally charged dipalmitoyl lecithin or lecithin (45 mol per cent), using these liposomes minimized morphological alterations and toxic reactions. The water-soluble nido-carborane anion-loaded PEGylated liposomes. PEGylated liposomes were utilized as boron carriers in efforts to reduce mononuclear phagocyte system (MPS) uptake and enhance drug delivery to tumours.After a single dose followed by 20 minutes of neutron irradiation, the boronated liposomes exhibited enormous tumour tissue penetration and localized within the tumour cell cytoplasm leading to complete tumour growth suppression (Lee *et al.,* 2020). In presence of high salt concentrations and a range of pH conditions, reconstituted membrane proteins in complex E. coli lipid vesicles to demonstrate controlled uptake of hydrophilic molecules (Mohanan *et al.,* 2020). The Drug delivery is focused on improving corneal adhesion and permeation by incorporating various bio-adhesive and penetration enhancing polymers. Liposomal photosensitizing agent, rostaporfin, was evaluated for the treatment of age-related macular degeneration. It is now under phase 3 clinical trial. Rostaporfin requires less frequent administration compared to verteporfin which reduced side effects. Investigators observed fourfold higher in vitro corneal flux from penicillin G-loaded SUVs. Liposomal formulation of TA produced two fold increase in drug concentration in both the cornea and aqueous humor in the rabbit model. The cytochrome-C (Cyt-C) loaded cationic liposomes were utilized for the treatment of selenite-induced cataract in rats. Fluconazole encapsulated liposomal formulation was evaluated in rabbit model having candidal keratitis and shown about 1.7 fold healing ability than direct fluconazole (Habib *et al.,* 2010). Due to bigger size (approx. 550 nm in size), liposomes endure at the site of injection and avoided lymphatic drainage. Also, cationic liposomes encapsulated higher amounts of liposome-bound low molecular weight heparin and released the drug in a fruitful manner, thus have longer retention time and stable concentration at the targeted site (Moon *et al.,* 2006).In rats implanted with 9L-gliosarcoma, accumulation of the CHP-coated liposome labelled with [14C]-DPPC increased by 4.5 times in the tumour and by 2.1 times in the ipsilateral brain than control liposome (Ochi *et al.,* 1990).The survival of 9L-glioma-implanted rats was investigated by the use of liposomes encapsulated with CDDP, c/s-platinum diamino dichloride which is an antitumor drug. Average survival was 35.3 days for the group treated with the CHP-coated liposomes in comparison to untreated cells whose survival rate is approx. 20 days. As respiratory drug delivery vehicle, liposomal aerosol has several advantages such as prevention of local irritation, improved stability in the large aqueous core with less toxicity and easy release (Fendler and Romero 1977). Currently approved antifungal saline-stabilized Amphotericin B liposomes namely Ambisome and Fungisome have a longer bioavailability, lower toxicity, and fewer adverse effects (Rudramurthy *et al.,* 2013). The only liposomal medication delivery system licenced for treating age-related macular degeneration is Visudyne, which works by preventing the growth of new blood vessels in the eye (Chang *et al.,* 2012). The viral envelope proteins seen in liposomal formulations known as virosomes are attached to their lipid membrane. Based on virosome technology, Epaxal and Inflexal V (Berna Biotech Ltd, Bern, Switzerland), two commercial vaccines for hepatitis A, are available now-a-days (Immordino *et al.,* 2006). In order to improve their in vitro and in vivo durability, COVID-19 mRNA-based vaccines recently used liposome protection. A liposome-based mRNA anti-COVID-19 vaccination developed by Pfizer/BioNTech and Moderna is currently being used all over the world. These vaccines were developed to support immunological responses while preserving the stability of liposomes in blood. The following four components made up the COVID-19 vaccines:Cholesterol and distearoylphosphatidyl choline, which are thought to be the major ingredients of traditional liposomes (Tai *et al.,* 2021), Cationic lipids, including 1,2-dioleoyl-3-(trimethylammonium) and propane (DOTAP), stabilise the particle, bind to the negatively charged mRNA, and phospholipids and cholesterol molecules provide the necessary structure. Consequently, mRNA encoding the Coronavirus spike protein would be contained in liposomes, engineered for stability in the bloodstream until internalised by phagocytic cells by endocytosis. The mRNA is then translated into the spike protein, eliciting an immune response that neutralises or eliminates the invading virus (Gregoriadis, *et al.,* 2021).

2. Encapsulation of Phytochemicals for Enhanced Drug Delivery

The challenges of using phytochemicals in cancer therapy and wound healing are due to their poor water solubility, stability, bioavailability, and target specificity. It emphasizes the development of nano-drug delivery systems, particularly liposomes, to enhance the therapeutic efficacy of these phytochemicals. Liposomes are biocompatible and biodegradable nanoparticles that can improve the stability, solubility, and bioavailability of phytochemicals, protect them from premature degradation, and enhance their target specificity to cancer cells. In Similipal Biosphere Reserve (SBR), major plant species include Centella asiatia, Nyctanthes arbortristis, Gymnema sylvestre, Rawolphia serpentine, Withania somnifera, Cucurma longa, Aloe vera *etc.* (Dikshit *et al.,* 2016). The phytochemical extracts like ethanolic extracts, methanolic extracts, aqueous extracts from these plants encapsulated within phytosomes and liposomes for targeted drug delivery, wound healing *etc.*. Also some lichen species like Dirinaria applanata and Parmotrema andium has phenols and flavonoids reported (Sahoo *et al.,* 2022). Secondary metabolites such as terpenoids, tannins, flavonoids, and phenolic acids, saponins were obtained from these medicinal plant species. This section explains encapsulation of several amphiphilic plant extracts as well as bioactive phytocompounds *viz.* curcumin, quercetin, resveratrol, and berberine in liposomes, demonstrating their enhanced anti-cancer activities in vitro and in vivo (Chavda *et al.,* 2022). Liposomes enhance the bioavailability of phytochemicals through several mechanisms:

1. **Improved Solubility**: Liposomes can encapsulate both hydrophilic and lipophilic phytochemicals, improving their solubility in aqueous environments, which is crucial for their absorption in the body. Encapsulating water-soluble phytochemicals in lipid bilayer of liposomes protects them from precipitation and improves their solubility in biological fluids (Gonclaves *et al.,* 2015).
2. **Protection from Degradation**: Liposomes or phytosomes protect phytochemicals from premature enzymatic degradation and metabolism in the gastrointestinal tract, liver, and other tissues, thereby prolonging their circulation time in the bloodstream. This barrier action of lipid bilayer preserves the integrity and stability of the phytochemicals (Thakur *et al.,* 2011, Isacchi *et al.,* 2017).
3. **Enhanced Absorption**: The lipid bilayer of liposomes can fuse with cell membranes, facilitating the direct delivery of phytochemicals into cells, which enhances their absorption and bioavailability (Dutt *et al.,* 2023)
4. **Sustained Release**: Liposomes can provide a controlled and sustained release of phytochemicals, maintaining therapeutic concentrations in the bloodstream for extended periods (Gordon *et al.,* 2012)
5. **Targeted Delivery**: Liposomes can be engineered for passive or active targeting to specific tissues or cells, such as cancer cells, which increases the local concentration of phytochemicals at the desired site of action and reduces systemic side effects (Sarwa *et al.,* 2015). Mucoadhesive polymer-based formulations can enable drug localisation, enhanced retention duration, and increased bioavailability (Zou *et al.,* 2014).

6. **Reduced Toxicity**: By encapsulating phytochemicals, liposomes can reduce the exposure of non-target tissues to the active compounds, thereby minimizing potential side effects and toxicity (Krishna *et al.*, 1996).

The pharmacokinetics and pharmacodynamics of water-insoluble, poorly bioavailable, and extremely toxic drugs were improved by the biocompatible, biodegradable, and low immunogenicity liposome formulation. To get beyond their early restrictions, liposomes underwent several changes to their composition and production method. More than 500 liposomal formulations are now undergoing various stages of clinical development, including some liposome formulations that have been approved for use in the market to treat several disorders.

Conclusion

This study highlights the significance of indigenous natural products in the SBR with potential therapeutic effects against infectious diseases. Artificial membrane models or liposomes, when integrated with taxonomy and various natural herbal products, underscore the significance of healing practices and plant utilization in bringing health practices to an advanced era. Numerous therapeutic applications have utilized liposomes. With the capacity of lengthy circulation residence durations, liposomes with increased medication delivery to targeted diseased areas are increasingly gaining clinical recognition. In addition, as compared to free complements, liposomal medications have lower toxicities and higher effectiveness. The surface modification of these liposomes with polyethylene glycol (PEG), polyacrylic acid (PAA), or poly-L-lactic acid (PLA) has enhanced bioavailability and stability. The creation of immunoliposomes and immunophytosomes, by the conjugation of monoclonal antibodies, fragments, or targeting moieties, has enhanced the active targeting of cancer cells. Even though liposomes are typically harmless and have minimal intrinsic toxicity, efforts should be made to investigate the environmental and toxicological effects of prolonged liposome exposure in both people and animals. In essence, liposomes have a vital role in contemporary delivery methods based on therapeutic uses and potential drug carriers.

Abbreviations

SUVs: Small Unilamellar vesicles; LUVs: Large Unilamellar vesicles; GUVs: Giant Unilamellar vesicles; MLV: Multilamellar vesicles; NBD: PE: (N: (7: Nitrobenz: 2: Oxa: 1,3: Diazol: 4: yl): 1,2: Dihexadecanoyl: sn: Glycero: 3: Phosphoethanolamine, Triethylammonium Salt); FACS: Fluorescence Activated Cell Sorter; DPPC: Dipalmitoylphosphatidylcholine; FDA: Food and Drug Administration; DOTAP: 1,2: dioleoyl: 3: (trimethylammonium) and propane; Rhod PE: 1,2: dioleoyl: sn: glycero: 3: phosphoethanolamine: N: (lissaminerhodamine B sulfonyl) (ammonium salt); SRB: Sulforhodamine; PrCFs: protein corona fingerprints; Cyt: C: Cytochrome C; MPS: mononuclear phagocyte system; PDAC: pancreatic ductal adenocarcinoma

REFERENCES

Akashi KI, Miyata H, Itoh H and KinositaJr K (1996) Preparation of giant liposomes in physiological conditions and their characterization under an optical microscope. Biophysical journal 71(6): 3242-50.

Akbarzadeh A, Rezaei-Sadabady R, Davaran S, Joo SW, Zarghami N, Hanifehpour Y, Samiei M, Kouhi M and Nejati-Koshki K (2013) Liposome: classification, preparation, and applications. Nanoscale research letters. 8(1): 1-9.

Allen TM and CullisPR (2013) Liposomal drug delivery systems: from concept to clinical applications. Advanced drug delivery reviews. 65(1): 36-48.

Angelova M and Dimitrov DS (1988) A mechanism of liposome electroformation. Trends in colloid and interface science, 59-67.

Bangham AD and Horne RW (1964) Negative staining of phospholipids and their structural modification by surface-active agents as observed in the electron microscope. J MolBiol 8: 660–668.

Bozzuto G and Molinari A (2015) Liposomes as nanomedical devices. International journal of nanomedicine.10: 975.

Bulbake U, Doppalapudi S, Kommineni N and Khan W (2017) Liposomal formulations in clinical use: an updated review. Pharmaceutics. 9(2): 12.

Caracciolo G, Pozzi D, Caminiti R and Amenitsch H (2005) Lipid mixing upon deoxyribonucleic acid-induced liposomes fusion investigated by synchrotron small-angle x-ray scattering. Applied Physics Letters.87(13): 133901.

Chang HI and Yeh MK (2012) Clinical development of liposome-based drugs: formulation, characterization, and therapeutic efficacy. International journal of nanomedicine.7: 49.

Digiacomo L, Giulimondi F, Pozzi D, Coppola A, La Vaccara V, Caputo D and Caracciolo G (2021) A proteomic study on the personalized protein corona of liposomes. Relevance for early diagnosis of pancreatic DUCTAL adenocarcinoma and biomarker detection.Journal of Nanotheranostics 2(2): 82-93.

Disalvo EA and BouchetAM. (2014) Electrophoretic mobility and zeta potential of liposomes due to arginine andpolyarginine adsorption. Colloids Surf A PhysicochemEng Asp. 440: 170–174.

Duzgunes N and Nir S (1999) Mechanisms and kinetics of liposome–cell interactions. Advanced drug delivery reviews. 40(1-2): 3-18.

Fanalista F, Birnie A, Maan R, Burla F, Charles K, Pawlik G, Deshpande S, Koenderink GH, Dogterom M and Dekker C (2019) Shape and size control of artificial cells for bottom-up biology. ACS nano.,13(5): 5439-50.

Fendler JH and Romero A (1977) Liposomes as drug carriers.Life Sciences,20(7): 1109-20.

Fenz, SF and Sengupta K (2012) Giant vesicles as cell models. Integr. Biol. 4, 982–995.

Furusato T, Horie F, Matsubayashi HT, Amikura K, Kuruma Y and Ueda T (2018) De Novo Synthesis of Basal Bacterial Cell Division Proteins FtsZ, FtsA, and ZipA inside Giant Vesicles. ACS Synth. Biol. 7 (4): 953–961.

Gregoriadis G (2021). Liposomes and mRNA: Two technologies together create a COVID-19 vaccine. Medicine in Drug Discovery. 12: 100104.

Habib FS, Fouad EA, Abdel–Rhaman MS and Fathalla D (2010) Liposomes as an ocular delivery system of fluconazole: in–vitro studies. Acta ophthalmologica 88(8): 901-4.

Himanshu A, Sitasharan P and Singhai AK (2011) Liposomes as drug carriers. IJPLS, 2(7): 945–951.

Immordino ML, Dosio F and Cattel L (2006) Stealth liposomes: review of the basic science, rationale, and clinical applications, existing and potential. International journal of nanomedicine.1(3): 297.

Jorgensen IL, Kemmer GC and Pomorski TG (2017) Membrane protein reconstitution into giant unilamellar vesicles: a review on current techniques. European Biophysics Journal.46(2): 103-19.

Karamdad K, Law RV, Seddon JM, Brooks NJ and Ces O (2015) Preparation and mechanical characterisation of giant unilamellar vesicles by a microfluidic method. Lab on a Chip. (2): 557-62.

Kataria S, Sandhu P, Bilandi AJ, Akanksha M and Kapoor B (2011) Stealth liposomes: a review. International journal of research in ayurvedaand pharmacy 1;2(5).

Kelly C, Jefferies C and Cryan SA (2011) Targeted liposomal drug delivery to monocytes and macrophages. Journal of drug delivery.

Kono K, Nakai R, Morimoto K and Takagishi T (1999) Temperature-dependent interaction of thermo-sensitive polymer-modified liposomes with CV1 cells. FEBS letters. 456(2): 306-10.

Lee KD, Nir S and Papahadjopoulos D (1993) Quantitative analysis of liposome-cell interactions in vitro: rate constants of binding and endocytosis with suspension and adherent J774 cells and human monocytes. Biochemistry.32(3): 889-99.

Lee W, Sarkar S, Ahn H, Kim JY, Lee YJ, Chang Y and Yoo J (2020) PEGylated liposome encapsulating nido-carborane showed significant tumor suppression in boron neutron capture therapy (BNCT). Biochemical and Biophysical Research Communications. 522(3): 669-75.

Lira RB, Dimova R and Riske KA (2014) Giant unilamellar vesicles formed by hybrid films of agarose and lipids display altered mechanical properties. Biophysical journal. 107(7): 1609-19.

Llu L and Yonetani T (1994) Preparation and characterization of liposome-encapsulated haemoglobin by a freeze-thaw method. Journal of microencapsulation. 11(4): 409-21.

Matsuki H, Kato K, Okamoto H, Yoshida S, Goto M, Tamai N and Kaneshina S (2017) Ligand partitioning into lipid bilayer membranes under high pressure: implication of variation in phase-transition temperatures. ChemPhys Lipids 209: 9-18.

Mohanan G, Nair KS, Nampoothiri KM and Bajaj H (2020) Engineering bio-mimicking functional vesicles with multiple compartments for quantifying molecular transport. Chemical science. 11(18): 4669-79.

Moon JW, Song YK, Jee JP, Kim CK, Choung HK and Hwang JM (2006) Effect of subconjunctivally injected, liposome-bound, low-molecular-weight heparin on the absorption rate of subconjunctival hemorrhage in rabbits. Investigative ophthalmology and visual science 47(9): 3968-74.

Morini MA, Sierra MB, Pedroni VI, Alarcon LM, Appignanesi GA and Disalvo EA (2015) Influence of temperature, anions and size distribution on the zeta potential of DMPC, DPPC and DMPE lipid vesicles. Colloids Surf B Biointerfaces 131: 54-58.

Motta I, Gohlke A, Adrien V, Li F, Gardavot H, Rothman JE and Pincet F (2015) Formation of giant unilamellarproteo-liposomes by osmotic shock. Langmuir 31(25): 7091-9.

Mulla TS, Thorat MS, Rayate Y and Nitalikar M (2019) Liposome as a drug carrier. Asian Journal of Research in Pharmaceutical Science 9(2): 141-7.

Ochi A, Shibata S, Mori K, Sato T and Sunamoto J (1990) Targeting chemo-therapy of brain tumor using liposome encapsulated cisplatin. Part 2. Pullulan anchored liposomes to target brain tumor. Drug Deliv Syst. 5: 261–271.

Ong SG, Ming LC, Lee KS and Yuen KH (2016) Influence of the encapsulation efficiency and size of liposome on the oral bioavailability of griseofulvin-loaded liposomes. Pharmaceutics 8(3): 25.

Palchetti S, Caputo D, Digiacomo L, Capriotti AL, Coppola R, Pozzi D and Caracciolo G (2019) Protein corona fingerprints of liposomes: New opportunities for targeted drug delivery and early detection in pancreatic cancer. Pharmaceutics 11(1): 31.

Parui PP, Sarakar Y, Majumder R, Das S, Yang H, Yasuhara K and Hirota S (2019) Chem. Sci.10: 9140-51.

Pradhan B, Kumar N, Saha S and Roy A (2015) Liposome: Method of preparation, advantages, evaluation and its application. Journal of Applied Pharmaceutical Research 3(3): 01-8.

Rudramurthy SM, Jatana M, Singh R and Chakrabarti A (2013) In vitro antifungal activity of Indian liposomal amphotericin B against clinical isolates of emerging species of yeast and moulds, and its comparison with amphotericin B deoxycholate, voriconazole, itraconazole and fluconazole. Mycoses. 56(1): 39-46.

Shah RK, Shum HC, Rowat AC, Lee D, Agresti JJ, Utada AS, Chu LY, Kim JW, Fernandez-Nieves A, Martinez CJ and Weitz DA (2008) Designer emulsions using microfluidics. Materials Today 11(4): 18-27.

Shi NQ and Qi XR (2021) Preparation of drug liposomes by reverse-phase evaporation.Liposome-Based Drug Delivery Systems: 37-46.

Shum HC, Lee D, Yoon I, Kodger T and Weitz DA (2008) Double emulsion templated monodisperse phospholipid vesicles. Langmuir 24(15): 7651-3.

Stachowiak JC, Richmond, DL, Li TH, Brochard-Wyart F and Fletcher DA (2009) Inkjet formation of unilamellar lipid vesicles for cell-like encapsulation. Lab Chip 9: 2003–2009.

Szoka Jr F and Papahadjopoulos D (1980) Comparative properties and methods of preparation of lipid vesicles (liposomes). Annual review of biophysics and bioengineering. 9(1): 467-508.

Tai TT, Wu TJ, Wu HD, Tsai YC, Wang HT, Wang AM, Shih SF and Chen YC (2021) A strategy to treat COVID–19 disease with targeted delivery of inhalable liposomal hydroxychloroquine: a preclinical pharmacokinetic study. Clinical and translational science. 14(1): 132-6.

Teh S-Y, Khnout, R, Fan H, and Lee AP (2011) Stable, biocompaticle lipid vesicle generation by solvent extraction-based droplet microfluidics. Biomicrofluidics 5: 044113.

Tsai F.-C, Stuhrmann B and Koenderink GH (2011) Encapsulation of active cytoskeletal protein networks in cell-sized liposomes. Langmuir 27.

Tsumoto K, Matsuo H, Tomita M and Yoshimura T (2009) Efficient formation of giant liposomes through the gentle hydration of phosphatidylcholine films doped with sugar. Colloids and Surfaces B: Biointerfaces. 68(1): 98-105.

Valenzuela SM (2007) Liposome techniques for synthesis of biomimetic lipid membranes. In Nanobiotechnology of biomimetic membranes. Springer, Boston, MA, 75-87.

vanSwaay D and DeMello A (2013) Microfluidic methods for forming liposomes. Lab on a Chip. 13(5): 752-67.

Walde P, Cosentino K, Engel H and Stano P (2010) Giant vesicles: preparations and applications. Chem Bio Chem. 11(7): 848-65.

Yang L, Gordon VD, Mishra A, Som A, Purdy KR, Davis MA, Tew GN and Wong GC (2007) Synthetic antimicrobial oligomers induce a composition-dependent topological transition in membranes. Journal of the American Chemical Society 129(40): 1

Chapter 14

Entomophagy and Entomotherapeutic Practices in India with Reference to Mayurbhanj District of Odisha

Lopamudra, Swagatika Tripathy, Mayuri Behera and Sukanta Kumar Nayak

Department of Biotechnology, Maharaja Sriram Chandra Bhanja Deo University, Takatpur, Baripada, Odisha
e-mail: sukantanayak@rediffmail.com

Introduction

Insects which rank among the top orders of animals in evolutionary success, comprise approximately 80 per cent of the total population of living organisms. Insects have been consumed as food and included in the diet of several societies worldwide since ancient times. Approximately 100 nations in Asia, Africa, and Latin America have long included insects as a fundamental part of their diets. These foods are highly nutritious and contain a significant amount of protein, making them a potential source of both food and protein. Insects are a great source of protein, calories, monounsaturated and polyunsaturated fats, several minerals and vitamins (including riboflavin, biotin, folic acid, and iron), and trace amounts of copper, iron, magnesium, selenium, phosphorus, and zinc.

Edible insects not only provide essential nutrients, but they also contain significant amounts of antioxidants such as phenols and flavonoids. Approximately 80 per cent of the global population incorporates insects into their dietary habits. A number of indigenous communities in India also use insects as a staple meal. Tribal people consume a range of notable edible insects, including caterpillars, grasshoppers, beetle grubs, termites, bees, ants (including red ants), and various aquatic insects. Insects are not only taken as a last resort during periods of food scarcity, but they are also intentionally included in the regular diet throughout the year (subject to availability) by many indigenous populations in India. In addition to food products, numerous bug species are utilised in traditional medicine and for medicinal purposes by various ethnic groups. Insects possessing ethno-medicinal properties have diverse applications in both humans and other animals, including aquaculture.

Bugs play an important role in our ecosystem and are necessary for our existence in many ways. In addition to sustenance, insects also provide humanity with numerous other beneficial resources. The most commonly utilised insect products by humans are honey and silk. Bees are responsible for producing over 1.2 million tonnes of commercial honey per year (FAO 2009), whereas silkworms produce over 90,000 tonnes of silk according to Yong-woo (1999). Edible insects have been a fundamental part of human diets since prehistoric times, however certain societies may exhibit hesitancy towards their consumption. The majority of edible insects are obtained from woods, and recently, several countries have started to develop advanced methods for mass-rearing them. They provide a significant opportunity to integrate traditional knowledge with contemporary scientific advancements in both affluent and developing countries, making them an essential element of any society's dietary practices.

Entomophagy

Entomophagy refers to the practice of humans eating insects. The name "entomophagy" is derived from the Greek terms "entomon" and "phagein," both of which imply "to eat." Throughout history, humans have ingested the eggs, larvae, pupae, and adult forms of several insects. There are almost 3000 ethnic groups who practise entomophagy. Approximately 80 per cent of countries worldwide incorporate a diverse range of 1000-2000 bug species into their diets. The FAO suggests that consuming insects has the potential to mitigate the ecological harm resulting from animal farming. The practice of humans consuming insects is referred to as entomophagy. The name "entomophagy" is derived from the Greek terms "entomon" and "phagein," both of which imply "to eat." Throughout history, humans have ingested the eggs, larvae, pupae, and adult forms of several insects. There are almost 3000 ethnic groups who practise entomophagy. Approximately 80 per cent of nations worldwide incorporate a diverse range of 1000-2000 bug species into their diets.

Since human entomophagy provides a cheap supply of high-protein food, it has significantly decreased the occurrence of protein shortages in poor societies. Given their successful adaptation to challenging habitats spanning more than 500 million years, insects have developed a diverse array of antimicrobial compounds to combat harmful microorganisms that pose a threat to their well-being. Consequently, insects have recently become the focus of research for their potential as a source of antibacterial and anti-cancer drugs. The UN Food and Agriculture Organisation is now formulating policy proposals to tackle the challenges of global food security by promoting the utilisation of insects

Entomophagy has a long history of involvement in human evolution, as stated by Fontaneto *et al.* (2011). Insects have historically played a substantial role in human nutrition, and literature from Asia, Africa, and Latin America has extensively recorded their nutritional contributions (Paoletti and Bukkens, 1997). Humans worldwide consume around 1900 species of insects, such as bees, grasshoppers, caterpillars, beetle grubs, winged termites, worms, ants, cicadas, and other aquatic insects (Bodenheimer, 1951). Within the South/Central Asia region, encompassing countries such as India, Pakistan, Nepal, and Sri Lanka, there exist a minimum of 52 edible species belonging to 45 genera, 26 families, and 10 orders (Gope and Prasad, 1983).

According to the 2004 report from the Food and Agricultural Organisation (FAO) of the United Nations, the larvae of several edible insects include significant amounts of vitamins, calcium, potassium, magnesium, zinc, and iron. Conversely, the chemical makeup and nutritional value might vary depending on factors such as geographical location, the specific plants that the insects consume, and the prevailing seasons and availability. Approximately 2 billion individuals belonging to over 300 ethnic groups across the globe regularly consume more than 1500 species of insects. India is no different in this regard (MacEvilly, 2000).

Nutritional Value of Edible Insects

Dietitians, medical experts, and doctors are interested in edible insects not only because of their taste but also their nutritional content. Several insects, such as certain caterpillars, ants, termites, and silkworms, are believed to possess health enhancing properties in addition to being highly nutritious. Insects are a rich protein source, with a substantial amount ranging from 25-75 per cent. Insects possess chitin in quantities ranging from 5 per cent to 20 per cent. Chitin is a polysaccharide that is present in the exoskeleton of certain insects. It constitutes around 6.71 - 15.98 per cent of the total carbohydrates found in these insects (Raksakantong, 2010). Scientists have discovered that numerous insects possess significant quantities of lipids. Therefore, they can be utilised to enhance lipid consumption.

Protein, the building block of all living things, is present in many essential physiological components. Insect body parts contain a substantial amount of protein. The protein content of over 100 edible insects, regardless of their life stage (egg, larva, pupa, or adult), typically ranges from 20 to 70 percent, as observed in a study conducted by Xiaoming *et al.,* in 2010. Ephemeroptera larvae include 66.26 percent of raw protein, Odonata larvae contain 40-65 percent, Homoptera larvae and eggs contain 40-57 percent, Hemiptera larvae contain 42-73 percent, Coleoptera larvae contain 23-66 percent, and Lepidoptera larvae contain 20-70 percent. Various analytical examinations have shown that insects possess a greater protein content compared to the majority of plants.

Fat not only provides energy but also shields and supports several organs in the body. Multiple publications assert that a significant number of edible insects possess a high fat content. The fat content of the adult stage is significantly reduced as compared to the larvae and pupal stages. Unlike animal fat, edible insects contain higher levels of fatty acids that are crucial for human health. Some examples of these organisms include the larvae and pupae of *Dendrolimus houi, Musca domestica, Chilo Fuscidentalis,* and several species of ants. Edible insects are rich in protein and fat while being low in carbs. Furthermore, research has revealed that insects possess substantial amounts of polysaccharide, which can enhance the human immune system (Xiaoming *et al.,* 2010). Edible insects contain a significant amount of chitin in their bodies and skin, with chitin levels ranging from 5 per cent to 15 per cent depending on the species. Further, edible insects contain significant amounts of trace elements such as iron (Fe), copper (Cu), sodium (Na), manganese (Mn), calcium (Ca), potassium (K), zinc (Zn), phosphorus (P), as well as vitamins A, carotene, B1, B2, B6, C, D, E, and K.

Table 14.1: Nutrient Profile of Various Edible Insects at different Stages of Growth

Insect Name	*Stage*	*Protein (per cent)*	*Moisture (per cent)*	*Fat (per cent)*	*Fibre (per cent)*	*Ash (per cent)*	*Ca (mg/kg)*	*P (mg/kg)*
Superworms	Larvae	19.7	57.9	17.7	2.7	1.0	177	2,370
Silkworms	Larvae	9.3	82.7	1.1	1.1	1.1	177	2,370
Waxworms	Larvae	14.1	58.5	24.9	3.4	0.6	243	1,950
House crickets	Adults	20.5	69.2	6.8	3.2	1.1	407	2,950
House crickets	Nymphs	15.4	77.1	3.3	2.2	1.1	275	2,520
Mealworms	Larvae	18.7	61.9	13.4	2.5	0.9	169	2,850
Soldier Flies	Larvae	17.5	61.2	14.0	3.0	3.5	9,340	3,560
Fruit Flies	Adults	21.0	69.1	5.9	2.2	3.1	526	4,080
Butterworms	Larvae	15.5	60.2	29.4	1.4	0.8	125	2,250
Six-spotted Cockroaches	Nymphs	18.8	50.8	26.8	1.0	—-	295	1,820

Status of Entomophagy in India

In India, several cultures have consumed insects since prehistoric times. Entomophagy has emerged as a more urgent concern in India in recent years. Several tribes in India consume approximately 255 types of insects as a dietary source. Sangma *et al.* (2016) found that Coleoptera ingested the highest proportion (34 per cent) of edible insect species, followed by Orthoptera (24 per cent), Hemiptera (17 per cent), Hymenoptera (10 per cent), Odonata (8 per cent), Lepidoptera (4 per cent), and Optra (2 per cent). Indigenous populations in at least 18 distinct Indian states and one union territory such as Assam, Arunachal Pradesh, Manipur, Nagaland, Meghalaya, Mizoram, Tripura, Sikkim, West Bengal, Odisha, Jharkhand, Chhattisgarh, Madhya Pradesh, Telangana, Kerala, Karnataka, Tamil Nadu, and the Andaman and Nicobar Islands have been observed engaging in the practice of entomophagy.

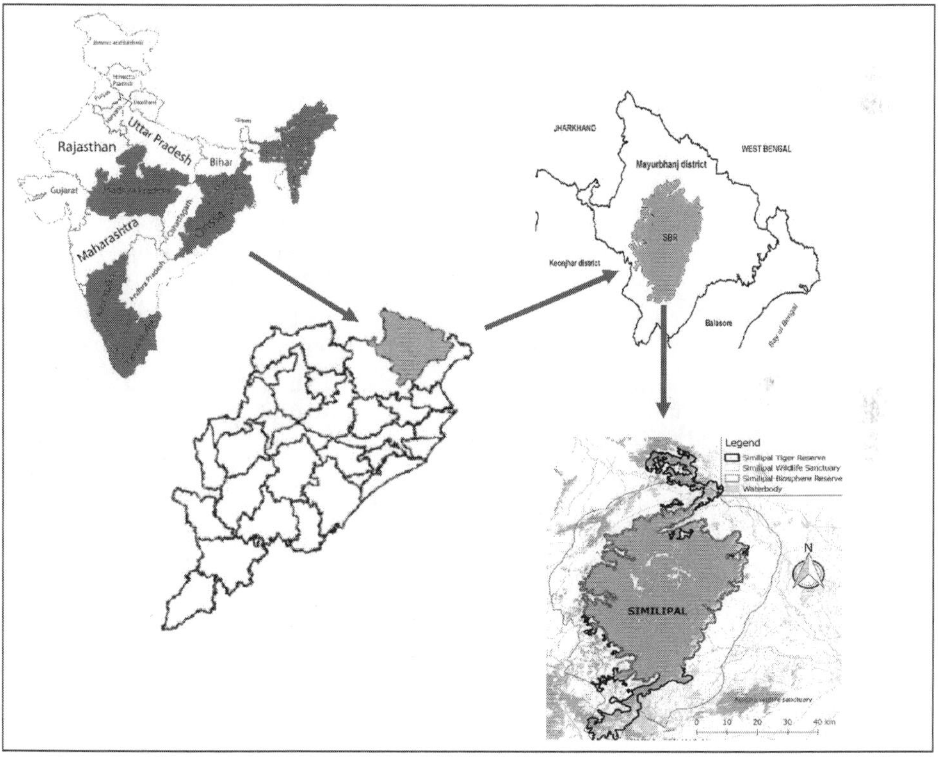

Figure 14.1: The Insect Consumption States of India with Reference to Mayurbhanj District of Odisha.

Northeastern Indian ethnic groups, especially those from Arunachal Pradesh, Assam, Manipur, and Nagaland, are known to eat insects. The indigenous communities of Meghalaya and Mizoram also consume insects to a certain degree. According to reports, the Mysore tribes and other indigenous people in the Carnatic region (now Tamil Nadu and southern Andhra Pradesh) devoured termites during the age of British conquest (Forbes, 1813). In another testimony from the early 20th century (Maxwell-Lefroy, 1909), it is documented that the indigenous peoples of South Indian presidencies also engage in the practice of entomophagy. People belonging to different tribes select edible insects based on their traditional customs, personal tastes, and the availability of these insects in their specific geographic location and time of year. Insects have long been utilised by numerous indigenous communities in Northeast India for sustenance, medicinal purposes, and various other applications.

In the Naga regions of Dimapur and Kohima, grasshoppers, crickets, red ants, and mulberry silkworm larvae are commonly eaten. The tribal inhabitants in the Midnapur district of West Bengal ingest red ant eggs and larvae. Various indigenous communities in Karnataka, Jharkhand, Tamil Nadu, and Kerela have a dietary preference for consuming red ants, winged termites, and bees as a source of sustenance (Srivastava *et al.,* 2009; Chakravorty, 2014). In Manipur, the local population predominantly consumes a high quantity of Hemipteran insects as their main source of sustenance. The indigenous communities of Arunachal Pradesh use caterpillars, pupae, and non-mulberry silkworms (Bombyx mori) into their customary cuisine (Gahukar, 2012).

Table 14.2: Different Types of Edible Insects Consumed in Various States of India

Scientific Name	*Common name/ Local name*	*Location*	*Types of uses*	*References*
Acheta domesticus	House Cricket/ Gusengra	Assam	Adult insects are fried	Hazarika *et al.* (2020)
Anomala spp	Shimphoo	Arunachal Pradesh	Roasted or fried	Sharma and Banu (2019)
Apis (*Cerana*) *indica*	Honey bee	Arunachal Pradesh	Immature stages are boiled or roasted.	Sharma and Banu (2019)
Apis dorsata	Rock bee/ Berema	Manas National Park, Assam	Eggs and larvae are consumed at raw	Hazarika *et al.* (2020)
Aristobia spp.	Chorgipok	Arunachal Pradesh	Roasted or fried	Sharma and Banu (2019)
Batocera roylei	Chorgipok	Arunachal Pradesh	Roasted or fried	Sharma and Banu (2019)
Brachytrupe orientalis	Loothong/ Idmon/Chur/ Cricket	Arunachal Pradesh	It is bound withlocal leaves androasted	Sharma and Banu (2019)
Brachytrupes spp.	Okul	Arunachal Pradesh	Adults are roasted/fried	Gogoi *et al.* (2021)

Scientific Name	*Common name/ Local name*	*Location*	*Types of uses*	*References*
Choroedocus robustus	Short-Horned Grasshopper/ Gumakhushep	Manas National Park, Assam	Adult insects are fried	Hazarika *et al.* (2020)
Diplonychus rusticus	Water beetle/ Amphu Dabla	Manas National Park, Assam	Adult insects are consumed either fried ora s curry.	Hazarika *et al.* (2020)
Dytiscus marginalis	Great diving beetle	Assam	Fried	Thangjam *et al.* (2020)
Eumenus spp.	Potter wasp	Northeast India	The "nest" of potter wasps to treat stomach problems.	Loganathan and Haldhar (2020)
Eupreponotus inflatus	Short-Horned Grasshopper/ Gumanargi	Manas National Park, Assam	Adult insects are fried or smoked	Hazarika *et al.* (2020)
Gryllotalpa spp.	Kuborr/Mole Cricket	Arunachal Pradesh	Adults are roasted/fried	Gogoi *et al.* (2021
Gryllus bimaculatus	Field Cricket/ Fendadangra	Manas National Park, Assam	Adult insects are fried or smoked before consumption.	Hazarika *et al.* (2020)
Gryllus spp	Cricket	Nagaland, Manipur, Assam, Arunachal Pradesh	Roasted	Thangjam *et al.* (2020)
Helicupris spp.	Shimpok	Arunachal Pradesh	Roasted or fried	Sharma and Banu (2019)
Ictinogomphus rapax	Dragonfly/ Gandula	Manas National Park, Assam	Nymphs are fried	Hazarika *et al.* (2020)
Laccotrephes ruber	Water scorpion/ Omabunda	Manas National Park, Assam	Adult insects are fried or smoked.	Hazarika *et al.* (2020)
Lepidiota spp.	Phobang	Arunachal Pradesh	Fried	Sharma and Banu (2019)
Lethocerus indicus	Giant Waterbug/ Gangjema	Manas National Park, Assam	Adult insects are fried or smoked.	Hazarika *et al.* (2020)
Lethocerus spp.	Kupthak/Giant Water bug	Arunachal Pradesh	Adults are roasted/fried	Gogoi *et al.* (2021)

Scientific Name	*Common name/ Local name*	*Location*	*Types of uses*	*References*
Odontotermes spp.	Queen termite	Karnataka	Consumed raw.	Chakravorty (2014)
Odontotermes spp.	Queen termite	Odisha	Raw termites or termites with rice	Chakravorty (2014)
Oecophyllas maragdina	Red ant	Odisha	Roasted insects with rice.	Srivastava *et al.* (2009)
Oecophyllas maragdina	Demta	Jharkhand	Eggs chutney	Srivastava *et al.* (2009)
Oecophyllas maragdina	Musuru red ants	Tamil Nadu	Adults, eggs, and larvae Rasam (Soup)	Sharma and Banu (2019)
Periplaneta americana	Cockroach	Northeast India	Consumed for the treatment of asthma problems.	Loganathan and Haldhar(2020)
Polistiso spp.	Nyasaa/Paper Wasp	Arunachal Pradesh	Eggs, larvae, pupae, and adults are boiled/fried	Gogoi *et al.* (2021)
Samia cynthia Ricini	Raijung	Arunachal Pradesh	Larvae and pupae are roasted/fried	Gogoi *et al.* (2021)
Vespa spp.	Hornet	Nagaland, Manipur, Arunachal Pradesh	Deep-fired	Thangjam *et al.* (2020)

Status of Entomophagy in Odisha Special Reference to Mayurbhanj District

Many indigenous peoples of Odisha have relied on insects for sustenance since ancient times. These include the Kandhas, Santals, Saura, Bondas, Kharia, Munda, and many more. The tribal inhabitants of several districts in Odisha, such as Kandhamal, Koraput, Sundergarh, Keonjhar, Rayagada, and Mayurbhanj (Srivastava *et al.*, 2009), consume red ants (Kai), flying termites (Jhari poka), house crickets (jhintika), date palm worm (Khajuri poka), and termites (uii). In certain regions of Odisha, individuals partake in the consumption of individuals partake in the consumption of chutney and acidic beverages made from the eggs, larvae, pupae, and adult forms of the red weaver ant. In addition to conventional cuisine, insects have been utilised as traditional treatments by indigenous communities since ancient times due to their potent medicinal properties, absence of negative side effects, convenient accessibility, and cost efficiency.

A total of 58.7 per cent of India's indigenous tribal peoples the Gond, Sabaras, Bhumija, Bhuyan, Bathudi, Kharia, Munda, Santhal, Kolha, Majhi, and Lodha live in the Mayurbhanj area. Throughout history, these people have traditionally consumed insects as a source of food. They hold the belief that these edible insects possess ethnomedicinal characteristics, which are considered to have the ability to treat various ailments such as asthma, cancer, and ulcers (Table 14.3). These ethnic communities are renowned in their local area for their expertise in utilising traditional medicine to treat a wide range of illnesses.

Table 14.3: Availability of different Types of Edible Insects and their Therapeutic Uses in the Mayurbhanj District of Odisha

Insects	*Local Name*	*Scientific Name*	*Body Parts Used*	*Disease*
Honey Bee	Mahu machhi	*Apis mellifara*	Whole body	Gastric Ulcer
Fire Fly	Julujulia poka	*Lampyridae*	Whole body	Stomach Ulcer
Red weaver ant	Kai	*Oecophylla smarangdina*	Whole body	Respiratory disease, Anti-inflammatory
Cockroach	Asarapa	*Blattaria*	Whole body	Asthma, Anti-inflammatory
Mole cricket	Kalikatra	*Scapteriscus didactylus*	Alimentary canal	Worm infection
Wasp	Birudi	*Vespula germanica*	Whole body	Gastric Ulcer
Termites	Uuii	*Odontotermes obesus*	Whole body	Bronchial problem and diuretic
House fly	Machhi	*Musca domestica*	Larvae	
Grasshopper	Jhintika	*Caelifera*	Whole body	Anti-inflammatory
Beetle	Gobara poka	*Coleoptera*	Whole body	Worm infection
Caterpillar	Silkworm/ Armyworm	*Papilio machaon*	Larvae	Stress reduction

Therapeutic Uses of Insects

Therapeutic insects can play a role in managing various challenges and promoting the overall health of any organisms. Many a times, for local remedies, one might look to traditional knowledge and the ethnic people's eating of insects as an example; however, this is only relevant for the traditionally living and somewhat "westernised" members of the population. Most individuals are unaware of the medicinal advantages that insects possess due to their status as a well-guarded secret. The Karbi tribe of Assam utilises Red grasshawk dragonflies (Neurothemis fluctuans) and Aeshna mixta dragonflies for the treatment of urinary issues in neonates.

Schistocerca gregaria, a species of desert locust, produces an oil that soothes cracked lips. Consuming the larva and pupa of the European fire ant (*Myrmica rubra*) has the potential to enhance human fertility. Apis mellifera indica honey is commonly utilised in culinary and cosmetic applications (Ronghang and Ahmed, 2010), as well as for the treatment of colds and coughs (Solanki and Chutia, 2008). In Nagaland, the eating of live mango borer (*Batocera titana*) larvae is used as a treatment for wounds. Insects possess several compounds with diverse biological properties, including antibacterial, diuretic, analgesic, anaesthetic, and antirheumatic effects (Yamakawa, 1998; Costa-Neto, 2005).

In Arunachal Pradesh, the ethnic communities utilise a mixture of roasted *Periplaneta americana* (Dictyoptera: Blattidae) extract and water as a remedy for tuberculosis and asthma. The powdered form of *Apis indica, A. mellifera,* and *A. florae* (Hymenoptera: Apidae) is mixed with honey to effectively treat conditions such as bolis, cough, and snakebite. According to Chakravorty *et al.* (2011), there is a belief among certain individuals that stomach ailments can be treated by consuming roasted or boiling nymphs of the mayfly species Ephemera danica. It's essential to note that the use of therapeutic insects requires careful consideration of the specific insect species, their impact on the target organisms, and the overall ecological balance within the system.

Table 14.4: Insects Used as Medicine for Treating Various Diseases

Local Name	*Scientific Name*	*Disease Treated*
House Fly	*Musca domestica*	Eye cysts, baldness
Locust	Anacridium sp.	Asthma, lung disease
Cicada	*Huechys sanguinea*	Ear infection
Cattle tick	*Boophilus microplus*	Chickenpox
Ground bettle	*Scarites spp.*	Healing of wound
Honey bee	*Apis spp.*	Cough, fever
Black ant	*Bothroponera rufipes*	Scabies, toothache
May fly	*Ephemera danica*	Stomach disturbances
Red velvet mite	*Trombidium grandisimum*	Pneumonia and fever
Weaver ant	*Oecophylla smargdina*	Stomach ache and dysentery
Silk moth	*Bombyx mori*	Pneumonia

Conclusion

Edible insects are not only a potential solution to global food security and environmental concerns but are also valued for their nutritional content. They are rich in essential amino acids, vitamins, and minerals. However, the acceptance of edible insects in Western cultures and some other regions can be influenced by cultural perceptions and the "yuck" factor associated with consuming insects. As awareness grows and more research is conducted, the use of edible insects may become more widespread. Since many Indian tribal communities are known to eat a substantial number of insects, entomophagy ought to be actively encouraged. Insects provide numerous benefits to humans due to their rich nutritional value, ability to reproduce quickly, low maintenance requirements, and ability to thrive on waste materials. To ensure the conservation of this invaluable natural asset and the indigenous knowledge for future cohorts, it is imperative to assess the diversity of insects and specifically the significance of ethno-entomophagy in India and Odisha. In order to guarantee the safety of food and feed, as well as the manufacturing of safe insect products, further research is necessary to create and streamline cost-effective, energy-efficient, and pathogen-free methods for growing, harvesting, and processing insects, as well as implementing sanitary processes. The use of insects is a dynamic and evolving field, and ongoing research is required to explore new possibilities and refine existing practices. It's important to conduct thorough assessments of each application, considering ecological, economic, and social factors for the sustainable integration of insects into animal and human systems.

REFERENCES

Alamu OT, Amao AO, Nwokedi CI, Oke OA and Lawa IO (2013). Diversity and nutritional status of edible insects in Nigeria: a review. International Journal of Biodiversity and Conservation 5: 215-222.

Bodenheimer FS. (c1951). Insects as Human Food: A Chapter of the Ecology of Man, The Hague: Dr. W. Junk P., Publisher.

Chakravorty J. (2014). Diversity of edible insects and practices of entomology in India: an overview. Biodiversity, Bioprospecting and Development. 1(3); 1000124

Doley AK., and J. Kalita. (2012). Traditional uses of insect and insect products in medicine and food by the mishing tribe of Dhemaji district, Assam, North-East India. Soc. Sci. and Res. 1; 11-21.

Fontaneto D., Tommaseo-Ponzetta M., Galli C., Risé P., Glew RH., Paoletti MG. (2011). Differences in fatty acid composition between aquatic and terrestrial insects used as food in human nutrition Ecology of Food and Nutrition. 50; 351-367.

Fontaneto D.,Tommaseo-Ponzetta M., Galli C., Risé P., Glew RH., Paoletti MG. (2011). Differences in fatty acid composition between aquatic and terrestrial insects used as food in human nutrition Ecology of Food and Nutrition. 50; 351-367.

Fromme A. (2002). Edible Insects.The Food Insects Newsletter; 8(2).

Gogoi R., Chetri S., and Ahmed R. (2021). Edible insects used as food by Tangsa and Wancho tribes of Changlang district, Arunachal Pradesh. Indian Journal of Entomology. e21036.

HazarikaAK., KalitaU., Khanna S., KalitaT., and Choudhury S. (2020). Diversity of edible insects in a Natural World Heritage Site of India: entomophagy attitudes and implications for food security in the region. Peer J. 8: e10248.

Kumar R. (2001). Insect pest of agriculture in Papua New Guinea Part-I: Principles and practice. pest of crops and stored products. UPNG Printery, Waigani. 723.

Maxwell-Lefroy H. (1909). Indian insect life; a manual of the insects of the plains (tropical India).

Meyer-Rochow VB. (2004). Ecological implications of mini-livestock; rodents, frogs, snails and insects for sustainable development. Science Publishers.

Meyer-Rochow VB. (2010). Entomophagy and its impact on world cultures: the need for a multidisciplinary approach. Edible forest insects. FAO Public: 23–37.

NC Jana., and PK Ghosh. (2015).Socio-Economic Conditions and Quality of Life in the Tribal Areas of Orissa with Special Reference to Mayurbhanj District. Space and Culture, India, Vol. 3; 25-41.

Paoletti MG., Bukkens SGF. (1997). Mini livestock. Ecology of Food and Nutrition. 36;95-346.

Sharma and Banu N. (2019). Entomophagy diversity in India a review. Journal of Emerging Technologies and Innovative Research. 6(1); 368- 374.

Srivastava SK., Babu N., and Pandey H. (2009).Traditional insect bioprospecting-as human food and medicine. Indian Journal of Traditional Knowledge. 8(4); 485-494

Thacker. Spink and Covan Huis A. (2019). Environmental sustainability of insects as human food. In Reference module in food science.

Thangjam R., Kadam V., Ningthoujam K., and Sorokhaibam M. (2020).A review on edible insects and their utilization in Northeastern Himalaya. Journal of Entomology and Zoology Studies. 8(3); 1309-1318.

Van HA., Van IJ., Klunder H., Mertens E., Halloran A., Muir G., Vantomme P. (c2013). Edible Insects: Future Prospects for Food and Feed Security. Food and Agriculture Organization of the United Nation.1-201.

VarshneyRK. (1997). Index *Rhopalocera indica*. Part III. Genera of Butterflies from India and neighbouring countries (Lepidoptera: C) *Lycaenidae.,*31; 83-138.

Yong-woo L (1999) Silk reeling and testing manual. FAO Agricultural Service Bulletin. 136.

Chapter 15

Biotechnological Advances in *Bombax ceiba* L.: A Historic Plant from the Similipal Biosphere Reserve, Odisha

Nihar Ranjan Singh

Department of Botany, Ravenshaw University, Cuttack–753003, Odisha

Introduction

Odisha is the coastal eastern state of India with a geographical land mass of approximately one hundred fifty-five thousand seven hundred seventeen (1,55,717) square kilometers and located between the coordinates of 17°47'N to 22°34'N latitude and 81°22'E to 87°29'E longitude. The geography of Odisha state is diverse with eastern ghats, the northern plateau region, and the coastal belt, which is connected with the ocean of the Bay of Bengal. This land mass also has major river like Mahanadi, Brahmani and Baitarani passing over and provide the necessary water resources for the vegetation in this part of India. As per the India State of Forest Report (ISFR) 2021, a biennial publication of Forest Survey of India (FSI) an organization under the Ministry of Environment Forest and Climate Change Government of India, Odisha state has around fifty-one thousand six hundred nineteen (51,619) square kilometers of land mass under forest vegetation which corresponds to close to 34 per cent of total Odisha state land mass. Taking into account the area under forest cover, the state of Odisha is below Madhya Pradesh, Arunachal Pradesh and Chhattisgarh.

Figure 15.1: Location of Similipal Biosphere Reserve, Odisha's First Biosphere Reserve.

The Simlipal biosphere reserve, Odisha's first biosphere reserve, is located in the Mayurbhanj district of Odisha with a geographical map of 20 ° 17 'N to 22 ° 34 'N latitude and 85 ° 40 'E to 87 ° 10 'E longitude. With a total land mass of five thousand five hundred sixty nine (5,569) square kilometers, Similipal biosphere reserve is one of the prominent biodiversity hotspot with rich floral diversity of Odisha (Rout *et al.,* 2022). Figure 15.1 gives an indication of geographical position of Similipal biosphere reserve. Similipal biosphere reserve is known for having predominant tree species like the sal (*Shorea robusta* Gaerten f.), jamun (*Syzygium cumini* (L.) Skeels), sahaju (*Terminaliaalata* Heyne ex Roth.), chiraoungi (*Buchanania lanzan* Spreng.). The name similipal is believed to be derived from the name of Simul tree, a historic plant belonging to the Similipal biosphere reserve (Behera, 2006).The Simul tree, also known as the Cotton tree belonging to the Bombacaeaefamily, isfound in many countries such assouthern China, India, Nepal, Africa, and Australia. The family Bombacaceae is closely related to the family Malvaceae and is represented in 17 genera with over 160 plant species. From an evolutionary perspective, the *Bombax* genus has prickles on its stem and branches and belonging to the *Kapok* clade that represent one of the most diverse lineages of the Bombacoideae (Malvaceae) (Carvalho-Sobrinho *et al.,* 2016). In the Indian subcontinent, there are three commonly seen plants belonging to Bombax genera, which are represented by *Bombax ceiba, Bombax insigne, Bombax scopuloram*. Out of which, *Bombax ceiba* is seen in the Similipal biosphere reserve.

Classification of *Bombax ceiba* L.

The *Bombax ceiba* L. tree is a fast-growing linear structure tree that grows in well-drained soils. The tree can reach up to 40 meters in height and can gain a girth of around 6 meters.

Figure 15.2: *Bombax ceiba* L., Plant Parts.
From clockwise – leaves, flower, fruits, and stem. (Representable images).

The leaves are digitate, glabrous with large compound with palmate appearance The woody region bore hard prickles, whereas the tree flowers are complete, bisexual, red coloured, large, and numerous with copious nectar. Fruits are edible and is capsule like with many black seeds (Chaudhary and Khadabadi, 2012). The taxonomic position of *Bombax ceiba* is as follows:

Kingdom: Plantae

Division: Magnioliophyta

Class: Magniolipsida

Order: Malvales

Family: Malvaceae (Bombacaceae)

Genus: *Bombax*

Species: *ceiba*

Binomial name: *Bombax ceiba* L.; *Bombax malabaricum* D.C.; *Salmalia malabarica* (D.C.) Schott and Endl.

Phytoconstituents and Traditional Therapeutics Use of *Bombax ceiba* L.

Traditionally, *Bombax ceiba* L. has been used for pharmacological purposes by tribal communities in Indian sub-continant. The flower and bark extracts of *Bombax ceiba* plant is given for 3 days to both men and women with sexual diseases like hydrocele, leucorrhoea, gonorrhea and is also used to check menstrual disorders in women by the tribal area. The people of Kandha tribe of Odisha, use the juice of fresh stem bark of the plant and mix with the extract of *Asparagus recemosus*, dried seeds of black peppers (*Piper nigrum* L.) and sugary stuffs to add taste for the cure of gonorrhea, impotency, spermatorrhea, sterility, nocturnal emission and leucorrhea. It is also prescribed to increase sperm in semen and acts as an aphrodisiac if taken several times for three weeks. It is also noted that the fresh paste prepared from the bark of *Bombax ceiba* mixed with cow dung is applied over back muscle of leg at night to treat inflammation. The seed paste of *Bombax ceiba* L. prepared in water was known to relieve small pox boils. In many tribal places, bark extract is externally used for treatment of cattle wounds. Bark juice is also used against dysentery, intestinal worms, and intestinal spasm when combined with *Mangifera indica juice* and *Psidium guajava* (Chaudhary and Khadabadi, 2012; Rajput, 2022; Ravi and Yogesh, 2019). Table 15.1 provides a list of different parts of *Bombax ceiba* plant and its extracts with known remedial effects.

Table 15.1: Evaluation of Therapeutic Properties from Extracts of *Bombax ceiba* L.

Sl.No.	*Plant Extract*	*Therapeutic Roles*	*References*
1	Methanolic extract of the *Bombax ceiba* L. roots	Phenolics ↑ and tannins ↑ (Radical scavenging activity)	(Jain *et al.*, 2011; Kriintong and Katisart, 2020)
2	Plant extracts (methanol and aqueous)	Against multidrug resistant *Salmonella typhii.*	(Rani and Khullar, 2004)
3	Methanolic extracts of *Bombax ceiba* L. leaves	Analgesic activity	(Sekar, 2015)
4	Aqueous and crude ethanol extracts of *Bombax ceiba* L. fruits	Diuretic Activity	(Jalalpure and Gadge, 2011)

Sl.No.	*Plant Extract*	*Therapeutic Roles*	*References*
5	Ethanolic extract of leaves of *Bombax ceiba* L.	Hepatoprotective	(Karole and Gupta, 2019)
6	Methanol extracts of *Bombax ceiba* L. stem barks	Antiangiogenic activity	(Rajput, 2022)
7	*Bombax ceiba* L. root extract	Aphrodisiac activity	(Khadabadi and Chaudhary, 2011; Rajput, 2022)
8	Water extracts of *Bombax ceiba* L. flowers	Anti-diabetic	(Yasien *et al.,* 2022)
9	Fractions of Bombax ceiba L.plant extracts	Immunomodulatory and anticancer activity	(Khaleel *et al.,* 2023; Sharma *et al.,* 2020)
10	Ethyl acetate and n-butanol fraction of the *Bombax ceiba* L. extract	Antidepressant	(Ojha *et al.,* 2021)
11	Water extracts of *Bombax ceiba* L. roots	Gain in body weight and sexual behaviour	(Bhargava *et al.,* 2012)

The plant *Bombax ceiba* L. possess wide range of phytochemical constituents. Using different solvent extraction methods, the plant parts like bark, leaves, flowers and fruits parts of *Bombax ceiba* L. yields different phytochemicals. The aqueous and ethanolic extracts of bark and leaves shows the presence of glycosides, phenols, flavonoids, carbohydrates and proteins (Karole *et al.,* 2018).The ethanolic extract of *Bombax ceiba* L. anther shows the presence of flavonoids, terpenoids, tannins, saponins and different phytochemicals whereas the aquatic extract indicated less quality of phytochemicals (Shukla *et al.,* 2020). Similarly, the leaves also contains many phytochemicals and many new metabolites are identified in recent times (Le *et al.,* 2022). Phytochemical investigations leads to isolation and identification of many diverse group of chemical compounds from different plants parts of *Bombax ceiba* L. (Refaat *et al.,* 2013). A selected list of phytoconstituents has been identified using different approaches and presented in the Table 15.2.

Recent Therapeutic and Medicinal Uses of *Bombax ceiba* L. Plant Extracts

Recent advances and technological progression have given researchers better methods to identify newer metabolites that can provide newer therapeutics applications and also provide beneficial role towards human health. In the following part, therapeutics role of plant extracts of *Bombax ceiba* L. has been evaluated against various diseases and shown potential beneficial role against disease progression.

Table 15.2: Identified Phytoconstituents from different Parts of *Bombax ceiba* L.

Plant Part	*Phytochemicals Identified*
Flower of *Bombax ceiba* L.	Apigenin, Cosmetin, Isovitexin, Linarin, Saponarin, Vicenin 2, Xanthomicrol, glucopyranoside, Kaempferol, Quercetin, glucosides, Daucosterol, Bombasin, Cedrol, Hentriacontanol, Neochlorogenic acid, Bombalin, Alanine, Glutamic acid, Glycocol, Arabinose, Galactose, Glucose, Lactose, Sucrose
Leaf of *Bombax ceiba* L.	Ceibanoside, Kaempferol, Naringenin, (+)-Epicatechin, ProcyanidinB2, Shamimin, Mangiferin
Root of *Bombax ceiba* L.	Hesperidin, Bombamalone, Lacinilene C, Hemigossylic acid, Isohemigossylic acid, Oleanolic acid, Triacontanol, Hemigossypolone, Gossypol, Hemigossypol, Carbolactone
Seed of *Bombax ceiba* L.	Gallic acid, Ethyl gallate, Tannic acid, Carotenoids, Hexacosanol, Vernolic acid, Octyl palmitate, Octadecyl palmitate, Cellulose, Pentosan, Polysaccharide consists of d-galactose and L-arabinose

Diabetes-mellitus

Different extracts of different plant parts of *Bombax ceiba* L., when consumed is known to downregulate the sugar concentration. Studies have shown that the presence of polyphenols, flavonoids, glycosaponins and polysaccharides are responsible for effective role towards management of diabetes. It is also established that the glucose uptake in to the cell is enhanced by 70 per cent in yeast cells in presence of *Bombax ceiba* aqueous flower extract (Yasien *et al.,* 2022). Apart from hypoglycemic property, the extracts of *Bombax ceiba* L. is shown to have hypolipidemic and hepatoprotective activities. Such extracts have shown significant role in preventing metabolic syndrome (Aly *et al.,* 2021). The administration of ethanolic extract of young root (400mg/kg) in mouse model have shown significant reduced blood glucose, higher High-density lipoprotein(HDL) level. It also decreases the level of SGOT and SGPT level, indicating the reduced hepatotoxicity in mouse model (Sharmin *et al.,* 2018). In another study, using ultra-performance liquid chromatography quadrupole-time-flight mass spectrometry (UPLC/Q-TOF-MS) techniques, the ethanolic extract from the flowers of *Bombax ceiba* L. identifies 24 phenolic acids and 6 flavonoids that have potential anti-diabetic roles. These compounds interferes with the intracellular receptor signaling pathways, unsaturated fatty acid, and DNA transcription pathways to regulate the sugar level in the cellular systems (Yin *et al.,* 2022)

Cancer

In recent times, plant-based metabolites are under investigation for their anti-cancer properties. Many plant-based compounds have shown huge potential of anti-cancer has been used in the treatment of cancer. Taxol, vincristine and vinblastine are few plant-based metabolites that are well known chemotherapeutic agents that target the microtubules of tumor cells. Similarly, many investigators have looked into several other plant groups for such metabolites and *Bombax ceiba* L. was also included. The methanolic extract of *Bombax ceiba* flowers have potential role in killing the cancerous cells in the mammary gland of the rats. The reduction of cancer-cell proliferation is indicated by the tumour markers as well as histopathology data (Mahmoud *et al.,* 2020). In one recent study, the effect of ethanol extract from *Bombax ceiba* flowers on three human cancer cells, lung A549, liver HepG2 and Huh7 cell lines were found to be effective. The proposed mechanism of such molecular genotoxicity effect is due to the presence of polyphenols that interfere with the DNA structure and production of reactive oxygen species that selectively kills the cancerous cells (Diab *et al.,* 2022)

Cardiomyopathy

Cardiovascular diseases are responsible for the highest death among many lifestyle diseases in the world. Cardioprotective role of plant extracts are well known and it is found that they can operate in addition with or without the commercial medications. *Bombax ceiba* L. is one among many other known plants with cardioprotective role as it has active phytochemicals in the form of flavonoids, polyphenols, plant sterol, plant sulphur compounds, and terpenoids. Flavonoids prevents low density lipoproteins (LDL), polyphenols are antioxidants, plant sterols decreases cholesterol absorption whereas plant sulphur compound prevents activation of nuclear factor-erythroid factor 2-related factor 2 (Nrf2) and inhibition of cholesterol synthesis. Terpenoids are known to decreasing atherosclerotic lesion in the aortic valve. All these phytochemicals are found in the *Bombax ceiba* L. plant extracts and thus makes it a suitable candidate for cardio protection medications (Bachheti *et al.,* 2022). In one study, the aqueous flower extract of *Bombax ceiba* L. was administered orally to Wister rats shown increase in the cardiac antioxidant enzymes and reduction in lipid peroxidation. Also, it is seen that the plant extract with vitamin E supplement can reduce mitochondrial swelling, leukocyte infiltration, lipid inclusions and myofibrillar loss (Patel *et al.,* 2011). In another study, administration of 3g of *Bombax ceiba* root powder significantly increased (up to 70 per cent) plasma fibrinolytic activity and up to (106 per cent) the serum antioxidant level in individuals with stable Ischemic Heart Disease (Jain *et al.,* 2012).

Commercial Applications

Humans have used plants for their needs. Commercial application of plant materials is playing vital role in the growth of a country economy. Presently, emphasis have been given on the generation of plant-based biofuel, plant-based nutraceuticals and plant-based fibres. A recent study has shown that the *Bombax ceiba* L. could be utilized as a potential lignocellulosic biomass source for bioethanol production. Alkali pre-treatment significantly exposed cellulose (60 per cent), which was further hydrolysed to fermentable sugars (50.9 per cent saccharification), which yielded better ethanol (54.51 g/L) production by *Saccharomyces cervisae* (Ghazanfar *et al.,* 2022).*Bombax ceiba* plant components are also known to produce nano particles. The flowers and seeds are used to synthesize nanoparticles (NPs) and that can be utilized in fuel production and blending or medicinal properties (Arun *et al.,* 2023; Safdar *et al.,* 2023). Another important product from the plant *Bombax ceiba* L. is the gum, also known as *moca-rasa*. Gum can be extracted from the dried calyx or leaves of *Bombax ceiba* L. and have been found to be use in Indian folk medicine, the ayurveda (Shah, 2022). Another applications of the flower of *Bombax ceiba* L. is used as a natural dye on cellulosic based textiles. It is seen that the linen fabric were dyed using flower extract shown better coloration properties (color values, color coordinates, and fastness properties) as well as functional properties *viz.* antibacterial activity, UV protection, and antioxidant activity (Sheikh *et al.,* 2019). On the other hand, the *Bombax ceiba* L. plant fibres are rich in naphthalene-based aromatic compounds and can interact with positively charged ions. Due to these factors, the fibres can adsorb pollutants like arsenic ions from aqueous streams. This shows the potential of *Bombax ceiba* L. fibres as pollutant reductant (Dindorkar *et al.,* 2022).The fibres can also be used in the making of mattress and cushions by the local people. The wood of the plant is also having commercial uses in the form of making of wooden boxes or for burning purposes. Apart from that the plant has ecological uses in the form of landscaping and forestation purpose.

Molecular Studies of *Bombax ceiba* L.

At molecular level, sequencing studies have been carried out recently that opens up plethora of information to the scientific community for genetic improvement and effective use of plant. The whole genome sequencing carried out on the intact genome shows the size of 895 Mb in length (Gao *et al.,* 2018). Apart from whole genome, the mitochondrial and chloroplast genome were also sequenced. The mitochondrial genome comprised of 5,94,390 base pairs with 35 protein coding genes. Along with that, the mitochondrial genome codes for 2 rRNA and 25 tRNA (Gao *et al.,* 2018). Contrary to mitochondrial genome, the chloroplast genome contains 1,58,997 base pairs. It has a total of 116 annotated genes that comprises of 81 protein coding genes, 8 rRNA and 27 tRNA genes (Gao *et al.,* 2018). In another study, the molecular origin of fiber from the fruits of *Bombax ceiba* L. has been carried out that provides necessary molecular methods to improve fiber yield and quality, so that it can be compared to cotton fiber. Using high-throughput single-cell RNA sequencing a total of 15,567 high-quality cells were identified from the inner wall of the *Bombax ceiba* L. ovary, and 347 potential marker genes for fiber initiation that can possibly provide information on the process of fiber initiation, development and differentiation (Ding *et al.,* 2023).

Discussion and Conclusion

Bombax ceiba L. is a long deciduous tree with beautiful flower that have great ecological and economic importance. A historical plant on which the Similipal biosphere reserve is possibly named indicates its significance. The plant is used as source for food, fodder, fuel, fiber and as traditional medicine in local folklore has been well known. Evaluation of phytoconstituents and their efficacy of plant parts and with different solvents has been done and evaluated with the finding that it possesses several groups of secondary metabolites. These metabolites are responsible for the beneficial role of plants and exploited by traditional healers for medicinal purposes. Recent molecular level studies also provided the mitochondrial, chloroplast and also the whole genome sequences that can be further used for improving various beneficial properties of the plant.

REFERENCES

Aly O, Elias TR, Agaibyi MNAMH, Rasheed WI, Yassen NN and Diab Y (2021) Antidiabetic and hepatoprotective activities of bombax ceiba extract in obese rats with metabolic syndrome. Plant Archives 21(1): 748-756.

Arun, S. B., Karthik, B. M., Yatish, K. V., Prashanth, K. N., and Balakrishna, G. R. (2023). Green synthesis of copper oxide nanoparticles using the Bombax ceiba plant: Biodiesel production and nano-additive to investigate diesel engine performance-emission characteristics. Energy, 274, 127345. https: //doi.org/ https: //doi.org/10.1016/j.energy.2023.127345

Bachheti, R. K., Worku, L. A., Gonfa, Y. H., Zebeaman, M., Deepti, Pandey, D. P., and Bachheti, A. (2022). Prevention and Treatment of Cardiovascular Diseases with Plant Phytochemicals: A Review. Evid Based Complement Alternat Med, 2022, 5741198. https: //doi.org/10.1155/2022/5741198

Behera, K. K. (2006). Ethnomedicinal plants used by the tribals of Similipal Bioreserve, Orissa, India: a pilot study. Ethnobotanical leaflets, 2006(1), 17.

Bhargava, C., Thakur, M., and Yadav, S. K. (2012). Effect of Bombax ceiba L. on spermatogenesis, sexual behaviour and erectile function in male rats. Andrologia, 44(s1), 474-478. https: //doi.org/https: //doi.org/10.1111/ j.1439-0272.2011.01210.x

Carvalho-Sobrinho, J. G., Alverson, W. S., Alcantara, S., Queiroz, L. P., Mota, A. C., and Baum, D. A. (2016). Revisiting the phylogeny of Bombacoideae (Malvaceae): Novel relationships, morphologically cohesive clades, and a new tribal classification based on multilocus phylogenetic analyses. Mol Phylogenet Evol, 101, 56-74. https: //doi.org/10.1016/j.ympev.2016.05.006

Chaudhary, P. H., and Khadabadi, S. S. (2012). Bombax ceiba Linn.: pharmacognosy, ethnobotany and phyto-pharmacology. Pharmacognosy Communications, 2(3), 2-9.

Diab, K. A., El-Shenawy, R., Helmy, N. M., and El-Toumy, S. A. (2022). Polyphenol Content, Antioxidant, Cytotoxic, and Genotoxic Activities of Bombax ceiba Flowers in Liver Cancer Cells Huh7. Asian Pac J Cancer Prev, 23(4), 1345-1350. https: //doi.org/10.31557/apjcp.2022.23.4.1345

Dindorkar, S. S., Kumar, V., and Patel, R. V. (2022). Adsorption of trivalent and pentavalent arsenic ions from the aqueous stream using natural Bombax ceiba fibers: Insights from density functional theory calculations. Total Environment Research Themes, 3-4, 100012. https: //doi.org/https: //doi.org/10.1016/j.totert.2022.100012

Ding, Y., Gao, W., Qin, Y., Li, X., Zhang, Z., Lai, W., Yang, Y., Guo, K., Li, P., Zhou, S., and Hu, H. (2023). Single-cell RNA landscape of the special fiber initiation process in Bombax ceiba. Plant Communications, 100554. https: //doi.org/https: //doi.org/10.1016/j.xplc.2023.100554

Gao, Y., Wang, H., Liu, C., Chu, H., Dai, D., Song, S., Yu, L., Han, L., Fu, Y., Tian, B., and Tang, L. (2018). De novo genome assembly of the red silk cotton tree (Bombax ceiba). GigaScience, 7(5). https: //doi.org/10.1093/gigascience/giy051

Gao, Y., Wang, H., Liu, C., Chu, H., Yan, Y., and Tang, L. (2018). The complete mitochondrial genome of Bombax ceiba. Mitochondrial DNA B Resour, 3(1), 313-314. https: //doi.org/10.1080/23802359.2018.1437814

Ghazanfar, M., Nadeem, M., Shakir, H. A., Khan, M., Ahmad, I., Franco, M., Chen, L., and Irfan, M. (2022). Valorization of Bombax ceiba Waste into Bioethanol Production through Separate Hydrolysis and Fermentation and Simultaneous Saccharification and Fermentation. Fermentation, 8(8), 386. https: //www.mdpi.com/2311-5637/8/8/386

Jain, V., Verma, S., and Katewa, S. (2012). Effect of Bombax ceiba root on some cardiovascular risk parameters in patients with ischemic heart disease. Asian J. Biol. Sci, 5, 351-357.

Jain, V., Verma, S., Katewa, S., Anandjiwala, S., and Singh, B. (2011). Free radical scavenging property of Bombax ceiba Linn. root. Res J Med Plant, 5(4), 462-470.

Jalalpure, S., and Gadge, N. (2011). Diuretic effects of young fruit extracts of Bombax ceiba L. in rats. Indian journal of pharmaceutical sciences, 73(3), 306.

Karole, S., Gautam, G., and Gupta, S. (2018). Physicochemical, Qualitative and Quantitative Phytochemical Analysis of the Leaf and Bark of Bombax Ceiba L (Red Silk Cotton Tree). Journal of Drug Delivery and Therapeutics, 8(6-s), 105-110.

Karole, S., and Gupta, G. (2019). Preparation and evaluation of phytosomes containing ethanolic extract of leaves of Bombax ceiba for hepatoprotective activity. Evaluation, 6(2), 1-5.

Khadabadi, Somashekhar and Chaudhary, Pankaj (2011). Aphrodisiac Activity of *Bombax ceiba* Linn. Extract in Male Mice. Pharmatutor.

Khaleel, A. M., Saleh, E. S., and Dosh, R. H. (2023). In Vitro Anticancer Activity of Different Fractions of Bombax Ceiba Plant Cultivated in Iraq That Compared with Vincristine as A Positive Control. HIV Nursing, 23(1), 536-541.

Kriintong, N., and Katisart, T. (2020). In vitro antioxidant and antidiabetic activities of leaf and flower extracts from Bombax ceiba. Pharmacognosy Research, 12(2).

Le, H.-T.-T., Duong, T.-H., Tran, Q.-T.-P., Pham, N.-K.-T., Nguyen, N.-H., Dinh, M.-H., Dang, V.-S., and Sichaem, J. (2022). A New Sesquiterpene Glycoside from Leaves of Bombax ceiba. Chemistry of Natural Compounds, 58(5), 853-856. https: //doi.org/10.1007/s10600-022-03815-6

Mahmoud, A. H., Metwally, N. S., Youness, E. R., El-Toukhy, S. E., Elmalt, H. A., and Al-Mokaddem, A. K. (2020). Antiproliferative Activity of Bombax ceiba Flower Extract against Mammary Gland Carcinoma in Rats. Syst. Rev. Pharm, 11, 1406-1415.

Ojha, A., Arora, R., Khan, R., Sah, S. P., and Upadhyaya, K. (2021). Bioactivity guided fractionation of Bombax ceiba L. root extract for antidepressant activity and elucidation of mechanism of action using in silico studies.

Patel, S. S., Verma, N. K., Rathore, B., Nayak, G., Singhai, A. K., and Singh, P. (2011). Cardioprotective effect of Bombax ceiba flowers against acute adriamycin-induced myocardial infarction in rats. Revista Brasileira de Farmacognosia, 21.

Rajput, R. T. (2022). Ethanomedicine and pharmacology of semal (Bombax ceiba L.)-A Indian medicinal plant: A review. Agricultural Reviews, 43(2), 145-153.

Rani, P., and Khullar, N. (2004). Antimicrobial evaluation of some medicinal plants for their anti–enteric potential against multi–drug resistant Salmonella typhi. Phytotherapy Research: An International Journal Devoted to Pharmacological and Toxicological Evaluation of Natural Product Derivatives, 18(8), 670-673.

Ravi, C., and Yogesh, S. (2019). Bombax ceiba linn.: a review.

Refaat, J., Desoky, S. Y., Ramadan, M. A., and Kamel, M. S. (2013). Bombacaceae: A phytochemical review. Pharmaceutical Biology, 51(1), 100-130.

Rout, D., Mohanta, M. R., and Sahu, S. C. (2022). Floristic diversity of climbing plants in tropical forests of Similipal Biosphere Reserve, Odisha, India. Notulae Scientia Biologicae, 14(1), 11003. https: //doi.org/10.15835/nsb14111003

Safdar, M., Aslam, S., Akram, M., Khaliq, A., Ahsan, S., Liaqat, A., Mirza, M., Waqas, M., and Qureshi, W. A. (2023). Bombax ceiba flower extract mediated synthesis of Se nanoparticles for antibacterial activity and urea detection. World J Microbiol Biotechnol, 39(3), 80. https: //doi.org/10.1007/s11274-022-03513-z

Sekar, M. (2015). Molecules of interest–mangiferin–a review. Annual research and review in biology, 307-320.

Shah, N. C. (2022). The identification, etymology and uses of Bombax ceiba (semal) sold by street vendors as Semarkanda: a review. Indian Journal of History of Science, 57(1), 10-15. https: //doi.org/10.1007/s43539-022-00027-3

Sharma, N., Kispotta, S., and Mazumder, P. M. (2020). Immunomodulatory and anticancer activity of Bombax ceiba Linn leaf extract. Asian Pacific Journal of Tropical Biomedicine, 10(9), 426.

Sharmin, R., Joarder, H., Alamgir, M., Mostofa, G., and Islam, M. (2018). Antidiabetic and hepatoprotective activities of Bombax ceiba young roots in alloxan-induced diabetic mice. J Nutrition Health Food Sci, 6, 1-7.

Sheikh, J., Singh, N., and Srivastava, M. (2019). Functional Dyeing of Cellulose-based (Linen) Fabric Using Bombax Ceiba (Kapok) Flower Extract. Fibers and Polymers, 20(2), 312-319. https: //doi.org/10.1007/s12221-019-8294-4

Shukla, R., Nandan, K., Shukla, A., and Kaur, A. (2020). Phytochemical analysis and nutritive value of Bombax ceiba Linn. (Petals). Plant Archives, 20(1), 1201-1206.

Yasien, S., Iqbal, M. M., Javed, M., Alnuwaiser, M. A., Iqbal, S., Mahmood, Q., Elkaeed, E. B., Dera, A. A., Alrbyawi, H., and Pashameah, R. A. (2022). Comparative Evaluation of Various Extraction Techniques for Secondary Metabolites from Bombax ceiba L. Flowering Plants along with In Vitro Anti-Diabetic Performance. Bioengineering, 9(10), 486.

Yasien, S., Iqbal, M. M., Javed, M., Alnuwaiser, M. A., Iqbal, S., Mahmood, Q., Elkaeed, E. B., Dera, A. A., Alrbyawi, H., Pashameah, R. A., Alzahrani, E., and Farouk, A.-E. (2022). Comparative Evaluation of Various Extraction Techniques for Secondary Metabolites from Bombax ceiba L. Flowering Plants along with In Vitro Anti-Diabetic Performance. Bioengineering, 9(10), 486. https: //www.mdpi.com/2306-5354/9/10/486

Yin, K., Yang, J., Wang, F., Wang, Z., Xiang, P., Xie, X., Sun, J., He, X., and Zhang, X. (2022). A preliminary study of the chemical composition and bioactivity of Bombax ceiba L. flower and its potential mechanism in treating type 2 diabetes mellitus using ultra-performance liquid chromatography quadrupole-time-flight mass spectrometry and network pharmacology analysis. Front Nutr, 9, 1018733. https: //doi.org/10.3389/fnut.2022.1018733

Chapter 16

Structure and Diversity of Tree Species in Tropical Dry Deciduous Forest of Similipal Biosphere Reserve, Odisha, India

Sabita Tudu, Mayanka M. Soren and Sudam C. Sahu

Department of Botany, Maharaja Sriram Chandra Bhanja Deo University, Sriram Chandra Vihar, Takatpur, Baripada–757003, Odisha
e-mail: sudamsahu.bdk@gmail.com

Introduction

Tropical forests are genetically enriched, widely productive, structurally very complex and basically considered as the treasure house of biodiversity (Roy *et al.*, 2002). In accordance with nature, the world's tropical forests alone are in charge of regulating the regional climate pattern and the global carbon cycle (Thomas and Baltzer, 2002). Tropical forests are often referred to be any forest cover that can exist and be developed in tropical and sub-tropical environments. Only 7 per cent of the earth's land surface is covered by tropical forests, but more than 50 per cent of all species on the planet are found there (Wilson, 1988). It is well known that tropical forests have produced a variety of vegetation types, and these vegetational variations are largely influenced by the local temperature range, photoperiod length, soil quality, and precipitation percentage. In general, the temperature about 23°C and the photo period (nearly 12 hours) of tropical forests are comparatively fixed but precipitations are always changeable. Thus, the inequality in plant formations in these forests is primarily due to the precipitation.

All over the world 52 per cent, out of the total forests are came under tropical forests and more than 42 per cent of the tropical forests are recognized as dry forests (Holdridge, 1967). The distinctive feature of tropical dry deciduous forests is their ability to develop and endure in a wide range of climatic circumstances, including 500–1500 mm of annual precipitation, 5-8 dry months, and nutrient-poor soil. In brief, it can be said that there are two bands of tropical dry deciduous forests where the drier area located between 10° and 20° N latitudes, and 10° and 20° S longitude of the sub-tropical desert, respectively (Moran, 2018). These types of vegetation are known as seasonally dry tropical forest, tropical dry deciduous forest, monsoon forests, catinga, cuabal (Singh and Chaturvedi, 2018). One of the most amazing adaptations of trees in tropical dry woods is their deciduousness. In the dry season, trees shed their leaves to prevent the greatest amount of water loss through the foliage. After the monsoon season has ended, the seeding procedure progressively progresses for 2 to 6 months. So that the trees must benefit from photosynthesis in the manner in which they typically do (Singh *et al.,* 2018).

Tropical dry woods are acknowledged as a global ecological zone (GEZ) by the Food and Agriculture Organization (FAO). The three continents that make up the world's tropical areas are Asia, Africa, and Latin America. According to Lamprecht (1989), Murphy and Lugo (1986), Africa has the world's greatest area of dry forests, but according to FAO (2003), Asia is home to 23 per cent of the world's tropical forests. According to FAO (2001), dry forests represent the drier types of forests like savannah (Africa), catinga and chaco (South America), and dry deciduous dipterocarp and woods (Asia). Hegner (1979) and Lampercht (1989) highlight their views on the four major forests areas in Asian continents as 1) the important species of *Shorea robusta* and *Tectona grandis* abound in the western dry monsoon zone of the Indian sub-continent, 2) the dry basin landscape of central Myanmar is the paradise of *Tectona* species and Dipterocarpaceae, and 3) the eastern forests of China and Japan are home to a variety of important species of trees, including the Khorat and Thailand's arid basin scenery are decorated by Dipterocarpaceae species. 4) Sunda Island's dry monsoon region.

Similipal Biosphere Reserve (SBR) is well known for its floral wealth and medicinal plants (Saxena and Brahmam, 1989). It includes tropical moist deciduous forests, dry deciduous hill forests, high level sal woods, grasslands, and savannas in addition to tropical semi-evergreen forests. There are few ecology and biodiversity studies are available in Similipal (Reddy *et al.,* 2007; Mishra *et al.,* 2007; Mishra *et al.,* 2012; Mohanta *et al.,* 2018). However, the tree diversity assessment on tropical dry deciduous forest stands of Badgaon forest range, SBR has not been conducted so far. Therefore, the current study aims to offer some strategic information about the structure and composition of the tropical dry deciduous forest of SBR, which may be useful to establish management techniques and conservation practices of these forests.

Materials and Methods

Study Area

The Similipal Biosphere Reserve (SBR) is located in the core of Mayurbhanj district of Odisha, India. It is located between 20°17" to 22°34" N and 85°40" to 87°10" E and run across the area of 5569sq km (Figure 16.1). The entire Biosphere Reserve is distributed among 3 major zones. They are core area (845 sq km), buffer zone (2129 sq km) and transitional zone (2595 sq km). The buffer zone (2129 sq km) is extended to 1905 sq km of Similipal tiger reserve, 77 sq km of Nato reserve forest and 147 sq km of Satkoshia reserve forest (Mohapatra *et al.,* 2008). Altitudinally, SBR ranges from 200 meters to 1162 meters. Similipal contributes 38 per cent to the total area of the Protected Area network in Odisha (Rout *et al.,* 2008). In 1956, it was officially selected as a tiger reserve and adopted in national conservation program "Project Tiger" in 1973 (Dash and Behera, 2013). Similipal is declared as wild life sanctuary in 1979 by the state government of Odisha. Initially, Similipal was famous for tiger reserve. Besides, tiger reserve it was proclaimed as biosphere reserve along with a transitional area of 2250 sq km in 1994 (Dash *et al.,* 2013). Among the recognized eighteen biosphere reserve, Similipal is designated as the sixth largest biosphere reserve in India. It is also an important biodiversity hotspot in Eastern India (Dash and Behera, 2012) and was enlisted world heritage site by UNESCO. The potential and usefulness of this region is tremendous.

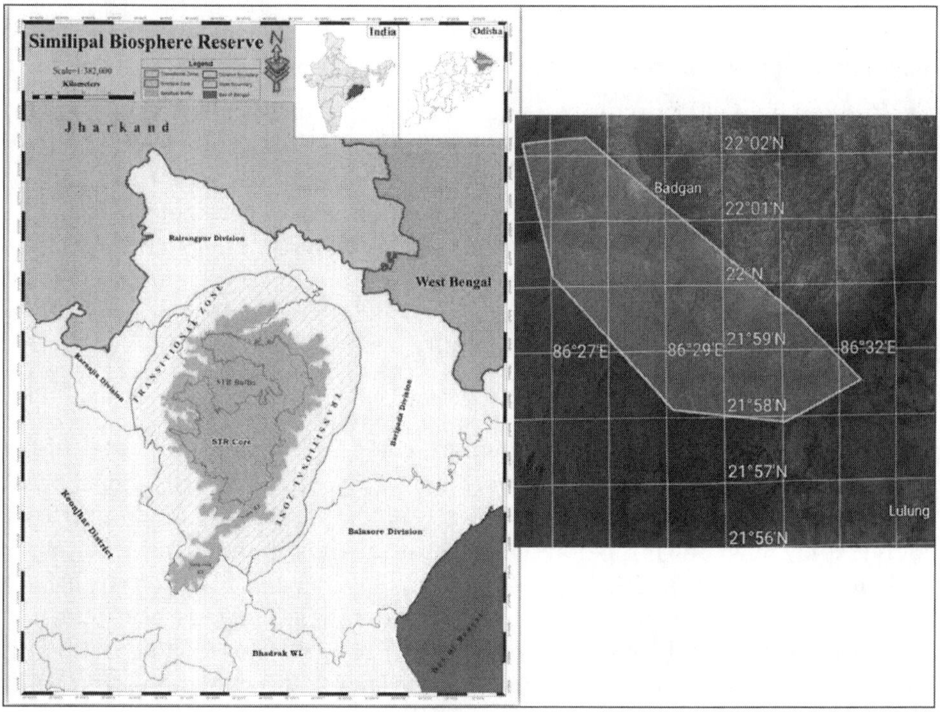

Figure 16.1: Map of the Study Area.

The area of SBR is surrounded by the plain of Baripada and Udala and extended to Joshipur in the north, Bisoi in the east and Thakurmunda in the west with numerous peaks, valleys, streams and rillets. The magnificent peaks of Khairiburu (1178 meters), Meghasini/Tunkiburu (1158 meters) and others make it attractive (Census, 2011). The hills avail well drainaged facility through numerous rivers, rivulets, streams and rillets. The perennial streams having watered in all directions link with main rivers like the Budhabalanga, the Subarnarekha and the Baitarani. The Budhabalang itself becomes a river where the khair, Bhandan, Birol, West deo, Salandi, Palpala and Khandakhai flow into the Subernarekha. All these rivers are flowing east and fall in to the Bay of Bengle. Two prominent waterfalls are Bareipani (400 meters), Joranda (150 meters) and other waterfalls like Devkund, Machhakandana and Sitakund are also well known for their beautiful surroundings.

Climate and Soil

India is a hot country so as the state Odisha too. There are three conspicuous seasons reign over in a year. They are the summer, rainy, and winter. The heat of the sun in summer is bearable and the temperature hardly goes above 40°C. The rainy season begins from mid-June to October month. The annual rainfall is about 2500mm which is occurred in some pockets and more in higher elevations inside Similipal biosphere reserve. The winter sits from mid-October to December. During the winter the temperature decreases to 5°C in part of the hill, forests, valleys and bare grasslands. The spring is pleasant. The temperature in the Similipal Biosphere Reserve is substantially influenced by topographic variance. In SBR, the southern and western regions are cooler, while the northern and eastern regions are warmer.

The soil type of a specific place is mostly determined by its geomorphology and old rock composition. The majority of Similipal is covered with red loam soil formed by Haematic rocks, which supports woody plantation (Ramakrishna *et al.,* 2006). Laterite soils and heavy clay soils are two more types of soil found on the plateau and throughout river basins, respectively, that help to support the reserve's different forest types. The soil of Similipal biosphere reserve is acidic and its pH varies from 4.8-6.8. Laterites developed in deep soil layer supports healthy vegetation growth (Ramakrishna *et al.,* 2008).

Field Sampling

Data on tree species diversity were obtained from various altitudes of the Badagaon forest. This study was conducted from November to February of 2020. The field sampling was carried out using the quadrant random sample approach. For this study, 50 m x 20 m quadrats were laid at 12 different sites in the tropical dry deciduous forest type of SBR. The size of the quadrat remains constant throughout the location. All plants were extensively surveyed, and the circumference of all tree species were measured with a measuring tape. The girth at breast height (GBH) of all trees was measured to be equal to or greater than 10cm. The Important Value Index (IVI) was calculated to know the tree species dominance. The geographic coordinates were obtained using the Global Positioning System (GPS). Identification of tree species was done with the help of regional floras (Saxena and Brahmam, 1994-96; Gamble and Fischer, 1915-35). Herbarium specimens were prepared and preserved in the Biosystematics laboratory, Department of Botany, North Orissa University, Baripada.

Data Analysis

Vegetation composition of the individuals recorded in the plots were evaluated by quantitative analysis *viz.* frequency, density, abundance using the formulas given by Curtis and McIntosh, 1950. The Importance Value Index (IVI) was used to rank each species and the plant species with the highest important value in the stand was considered the dominant species. It was calculated by using the formula given by Phillips, 1959. To represent the structural composition of the site, girth class analysis was also carried out. For an evaluation of the richness and variety of the tree species in all sampled plots, the distribution data were utilized to generate diversity indices such as Shannon diversity index and Simpson's index.

Distribution of tree species was calculated using the ratio of abundance to frequency (Whitford, 1949). If the value lies between <0.025: regular, 0.025-0.05: random and if >0.05 contiguous (Curtis and Cottam, 1956). The family importance value (FIV) was computed by following the equation given by Mori *et al.* (1983).

Results

From the study area, 706 individuals altogether, representing 41 species, 38 genera, and 26 families, were documented. The most diversified family in terms of species was Rubiaceae (6 species), which was followed by Euphorbiaceae (5 species), Fabaceae (3 species), Anacardiaceae, Combretaceae, Ebenaceae, and Salicaceae (all of which had 2 species), while the remaining 19 families were each represented by a single species (Table 16.2). At the genus level *Terminalia* and *Diospyros* had maximum 2 species *i.e.*, *Terminalia alata* Heyne ex Roth., *Terminalia chebula* Retz., and *Diospyros melanoxylon* Roxb., *Diospyros sylvatica* Roxb., respectively. Whereas, the other genus had represented only single species each. The ratio of species to genera (1.07) was less than the of ratio genus to family (1.5). The dominant tree species in descending order of IVI were *Shorea robusta* Gaertn.f. (65.47), *Croton persimilis* Mull.Arg. (18.51), *Terminalia alata* (15.32), *Protium serratum* (Wall. ex Colebr.) Engl. (13.86), *Buchanania lanzan* Spreng. (13.75), *Anogeissus latifolia* (Roxb. ex DC.) Wall. ex Guill and Perr. (12.50), *Cleistanthus collinus* (Roxb.) Benth. ex. Hook.f. (12.38), *Bridelia retusa* (L.) Spreng. (11.79), *Madhuca indica Gmel.* (10.69), and *Cassia fistula* L. (9.77) (Table 16.1). Out of all the species, the IVI of top ten dominant species contributed 61.36 per cent.

On the basis of contribution of basal area of different tree species, the total basal area of *Shorea robusta* (5.31 m^2/ha.) is at the top which was followed by *Protium serratum* (1.4 m^2/ha.), *Croton persimilis* (1.25 m^2/ha.), *Madhuca indica* (1.21m^2/ha.) and *Buchanania lanzan* (1.06 m^2/ha.). These five tree species made up 49.25

per cent (10.23 m^2/ha) of the study's overall basal area (Table 16.2). The diversity indices determined by Shannon (H) and Simpson were 2.056 ± 0.740 and 0.742 ± 0.231, respectively. The Shannon and Simson indices range from 0.892 to 2.757 and 0.357 to 0.918, respectively. The area's evenness index was calculated to be 0.593 ± 0.214. In the total study, 28 species were found with contiguous distribution, 11 species with random and 2 species with regular distribution (Figure 16.2).

Table 1. List of tree species and their density (Ind./ha), Frequency (per cent), basal area (m^2/ha) and IVI in Badagaon forest range, Similipal Biosphere Reserve, Odisha

Sl. No.	*Name of the Tree*	*Family*	*F (per cent)*	*D (Ind./ ha)*	*Do (m^2/. ha)*	*IVI*
1	*Alangium salviifolium* (L.f.) Wangerin	Alangiaceae	16.66	5	0.08	2.37
2	*Albizia lebbeck* (L) Corr.	Mimosaceae	16.66	2.5	0.26	2.82
3	*Anogeissus latifolia* (Roxb. ex DC.) Wall. Ex Guill & Perr.	Fabaceae	66.66	19.16	1	12.5
4	*Antidesma acidum* Retz.	Phyllanthaceae	8.33	0.83	0.03	0.88
5	*Bridelia retusa* (L.) Spreng.	Euphorbiaceae	75	21.66	0.65	11.79
6	*Buchanania lanzan* Spreng.	Anacardiaceae	75	21.66	1.06	13.75
7	*Caeseria tomentosa* Roxb.	Salicaceae	8.33	0.83	0.18	1.6
8	*Careya arborea* Roxb.	Lecythidaceae	75	11.66	0.48	9.25
9	*Casearia graveolens* Dalz.	Salicaceae	41.66	5	0.09	4.03
10	*Cassia fistula* L.	Caesalpiniaceae	83.33	14.16	0.38	9.77
11	*Cleistanthus collinus* (Roxb.) Benth. ex. Hook.f.	Euphorbiaceae	58.33	31.66	0.65	12.38
12	*Croton persimilis* Mull. Arg.	Euphorbiaceae	66.66	47.5	1.25	18.51
13	*Dillenia pentagyna* Roxb.	Dilleniaceae	8.33	2.5	0.05	1.24
14	*Diospyros melanoxylon* Roxb.	Ebenaceae	50	8.33	0.56	7.44
15	*Diospyros sylvatica* Roxb.	Ebenaceae	25	5	0.19	3.44
16	*Ehretia laevis* Roxb.	Boraginaceae	8.33	1.66	0.08	1.23
17	*Ficus benghalensis* L.	Moraceae	8.33	0.83	0.01	0.78
18	*Gardenia gumifera* L.f.	Rubiaceae	8.33	0.83	0.18	1.56
19	*Haldina cordifolia.* (Roxb.) Ridsd.	Rubiaceae	58.33	16.66	0.56	9.41
20	*Holarrhena pubescens* Andr.	Apocyanaceae	25	4.16	0.04	2.55
21	*Ixora pavetta* Andr.	Rubiaceae	25	4.16	0.36	4.11
22	*Lannea coromandelica* (Houtt.) Merr.	Anacardiaceae	41.66	8.33	0.19	5.08

Sl. No.	*Name of the Tree*	*Family*	*F (per cent)*	*D (Ind./ ha)*	*Do (m^2/. ha)*	*IVI*
23	*Madhuca indica* Gmel.	Sapotaceae	50	9.16	1.21	10.69
24	*Mallotus philipensis* (Lam.) Muell. Arg.	Euphorbiaceae	16.66	6.66	0.18	3.14
25	*Maytenus bailadillana* (V. Naray. and Mooey) D.C.S. Raju and Babu	Celastraceae	16.66	2.5	0.02	1.62
26	*Meyna spinosa* Roxb. Ex Link	Rubiaceae	33.33	6.66	0.2	4.3
27	*Miliusavelutina* (A.D.C.) Nook.f. & Thomson	Annonaceae	25	5	0.11	3.03
28	*Mitragyna parvifolia* (Roxb.) Korth.	Rubiaceae	16.66	1.66	0.09	1.85
29	*Oroxylum indicum* (L.) Benth. ex Kurz.	Bignoniaceae	33.33	3.33	0.09	3.23
30	*Phyllanthus emblica* L.	Euphorbiaceae	8.33	0.83	0.01	0.77
31	*Protium serratum* (Wall.ex Colebr.) Engl.	Buraseraceae	58.33	19.16	1.4	13.86
32	*Pterocarpus marsupium* Roxb.	Fabaceae	50	14.16	0.52	8.21
33	*Schleichera oleosa* (Lour.) Oken	Sapindaceae	41.66	13.33	0.82	8.98
34	*Shorea robusta* Gaertn.f.	Dipterocar-paceae	91.66	199.16	5.31	65.47
35	*Syzygium cumini* (L.) Skeels.	Mytraceae	66.66	14.16	0.45	8.99
36	*Terminalia alata* Heyne ex Roth.	Combretaceae	83.33	33.33	0.86	15.32
37	*Terminalia chebula* Retz.	Combretaceae	25	4.16	0.1	2.88
38	*Vitex peduncularis* Wall.ex Schauer.	Lamiaceae	8.33	1.66	0.01	0.92
39	*Wendlandia tinctoria* (Roxb.) DC.	Rubiaceae	8.33	2.5	0.07	1.33
40	*Xylia xylocarpa* (Roxb.) Taub.	Fabaceae	25	15.83	0.64	7.46
41	*Ziziphus glabrata* (Retz.) Wild.	Rhamnaceae	8.33	0.83	0.11	1.25
	Grand Total		1516.66	588.33	20.77	

F: Frequency, D: Density, BA: Basal area, IVI: Importance Value Index.

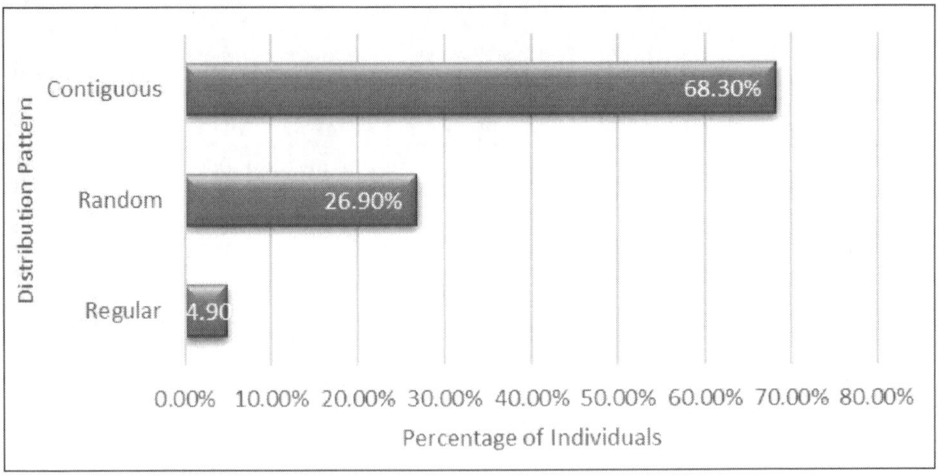

Figure 16.2: Distribution Pattern of the Tree Species in Badagaon Forest Range, Similipal.

In the study area, there were a wide variety of tree species with respect to abundance. According to Pragasan and Parthasarathy (2010), tree species were divided into five categories based on their abundance: very rare (1-2 individuals), rare (3-20 individuals), common (21-100 individuals), dominants (101- 500 individuals) and predominant (> 501 individuals). On analysis it was found that 9 species were very rare, 24 were rare, 7 species were common and only one species came under dominant category. There was no single predominant species out of all. Dominant, common, rare and very rare species contributed 33.852 per cent (239 individuals), 33.002 per cent (233 individuals), 31.444 per cent (222 individuals) and 3.116 per cent (12 individuals), respectively to the forest composition.

In the study area, species richness generally reduces as girth class increases. Among 41 tree species listed, maximum number of individuals (418) were recorded in the lowest girth class(10-30cm) followed by 195, 48,22, 11, 5, 4 and 3 for GBH class 31-50 cm, 51-70 cm, 71-90 cm,91-110 cm, 111-130 cm, 131-150 cm, and more than 150 cm, respectively. It was observed that trees of lower girth class (GBH of 10-30cm and 31-50cm) represented maximum number and contributed to 86.82 per cent (618 individuals) to the total no of individuals (706).

According to the calculated FIV (Family Index Value), Dipterocarpaceae had the highest FIV value (92.29) followed by Euphorbiaceae (49.68), Fabaceae (26.79), Rubiaceae (17.97), Combretaceae (17.26), Anacardiaceae (16.01), Buraseraceae (13.01) and FIV values for other 19 families have less than ten (< 10). The families with more than 10 FIV value have contributed 77.95 per cent to the total FIV values of all the families found in the study area. On the hand, basal area of different families Dipterocarpaceae (6.37m^2) is dominant followed by Euphorbiaceae (3.33m^2), Fabaceae (2.61m^2), Rubiaceae (1.78m^2), Buraseraceae (1.68m^2), Anacardiacea

(1.5 m^2), Sapotaceae (1.45 m^2), Mimosaceae (1.3m^2), and Combretaceae (1.17m^2). Other 17 families have FIV less than 1 (Table 16.2).

Table 16.2: Family Composition and FIV of Tree Species in Badagaon Forest Range, Similipal

Sl. No.	Family	No of Genera	No of Species	No of Individuals	Total Basal Area	R. Basal Area	R. Diversity	R. Density	FIV
1	Alangiaceae	1	1	6	0.1	0.84	0.84	0.84	2.54
2	Mimosaceae	1	1	3	1.32	5.1	0.42	0.42	5.95
3	Fabaceae	3	3	59	2.61	10.08	8.35	8.35	26.79
4	Phyllanthaceae	1	1	1	0.04	0.18	0.14	0.14	0.46
5	Euphorbiacea	5	5	130	3.33	12.85	18.41	18.41	49.68
6	Anacardiaceae	2	2	36	1.5	5.81	5.09	5.09	16.01
7	Salicaceae	2	2	7	0.33	1.29	0.99	0.99	3.28
8	Lecythidaceae	1	1	14	0.58	2.24	1.98	1.98	6.2
9	Caesalpiniaceae	1	1	17	0.46	1.79	2.4	2.4	6.61
10	Dilleniaceae	1	1	3	0.06	0.26	0.42	0.42	1.11
11	Ebenaceae	1	2	16	0.91	3.54	2.26	2.26	8.07
12	Boraginaceae	1	1	2	0.09	0.38	0.28	0.28	0.95
13	Moraceae	1	1	1	0.02	0.08	0.14	0.14	0.36
14	Rubiaceae	6	6	39	1.78	6.89	5.52	5.52	17.94
15	Apocynaceae	1	1	5	0.04	0.18	0.7	0.7	1.6
16	Sapotaceae	1	1	11	1.45	5.61	0.17	1.55	7.34
17	Celastraceae	1	1	3	0.02	0.1	0.42	0.42	0.95
18	Annonaceae	1	1	6	0.13	0.51	0.84	0.84	2.21
19	Bignoniaceae	1	1	4	0.11	0.45	0.56	0.56	1.58
20	Buraseraceae	1	1	23	1.68	6.49	3.25	3.25	13.01
21	Sapindaceae	1	1	16	0.98	3.81	2.26	2.26	8.34
22	Dipterocarpaceae	1	1	239	6.37	24.58	33.85	33.85	92.29
23	Myrtaceae	1	1	17	0.54	2.1	2.4	2.4	6.92
24	Combretaceae	1	2	45	1.17	4.51	6.37	6.37	17.26
25	Lamiaceae	1	1	2	0.02	0.09	0.28	0.28	0.65
26	Rhamnaceae	1	1	1	0.14	0.54	0.14	0.14	0.82

Discussion

It is generally accepted that trees support the lives of nearly all other forest species by providing habitat, resources, and food. Accordingly, the diversity of different kinds of trees has been recognized by scientists as essential to tropical forest biodiversity (Huston, 1994; Whitmore, 1998). The research investigation indicated 41 tree species, which is very few compared to the Tropical Dry Forest of Bannerghatta National Park in the Eastern Ghats, Southern India (128 species) (Gopalkrishna *et al.*, 2016), but very comparable to the Malyagiri Hills (57 species) of the Eastern Ghats (Sahu *et al.*, 2012). The variation in these forests' species counts is primarily caused by variations in the sample sites and degree of disturbance. While comparison sites were substantially larger than the present.

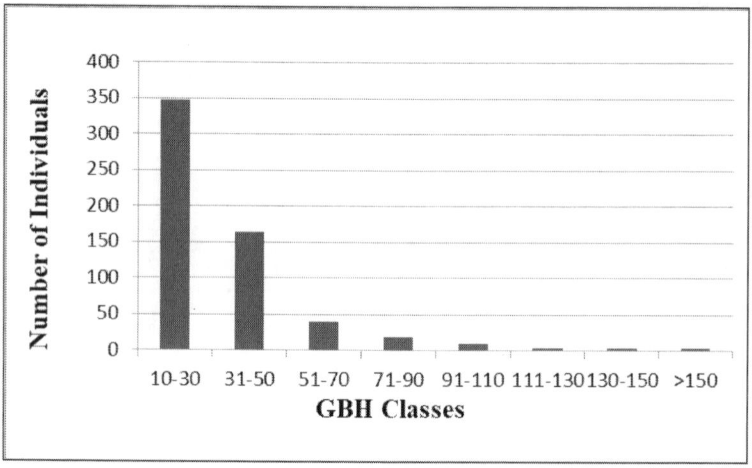

Figure 16.3: Distribution of Number of Individuals of Trees in different GBH Classes at Badagaon Forest Range, SBR.

The current investigation was conducted within 1.2 hectares of the sampling area. However, a number of factors (climatic and topographic) have been shown to affect the variety and abundance of species of forest vegetation. In general, the structure and content of the core area of the SBR are denser than those of the transition areas. The current study location is in a transitional area where there is direct human-forest contact. There were often visible signs of tree lopping and cutting in the neighbourhood. The local population frequently depends on outside sources for food, firewood, home construction, and agricultural equipment. The functional variety, ecological processes, and ecosystem services of the forest are greatly influenced by its density. The average stand density of the Badagaon forest range in Similipal (588.33 stems/ha) is within the range reported for other tropical forests on the Indian subcontinent, including Mudumalai forest reserve (298 stems/ha) (Sukumar *et al.*, 1992), Sinharaja (689 stems/ha), and Sri Lanka (Condit *et al.*, 2000). The proportion of trees in the various GBH classes determines how well and productively a forest works. According to several studies, a typical dry forest may support a maximum density of trees in the lowest girth classes (between 10 and 60 cm GBH) and a sizable number of trees in the higher girth classes (> 100 cm GBH) (Sahu *et al.*, 2012; Reddy *et al.*, 2007). However, in the current study, the lower GBH class contained more than 50 per cent of the entire population, while the higher GBH class contained an insignificant proportion of the population (Figure 16.2). The high level of anthropogenic disruptions in the area is what has led to the current scenario. According to Denslow (1995), Hett and Loucks (1976), and Poorter *et al.* (1996), the diameter distribution is a particularly useful tool for spotting trends in regeneration patterns. Beyond 10–30 cm, the stem density and species richness are reduced. The level of forest disturbance is revealed by the low diversity and density of trees with greater girth classes and the high diversity and density of trees with lower GBH seen in this study. Studying the distribution of tree densities across various girth classes made it simple to observe how the site's resources were being used. It is assumed that tree crops are not using forest land to its maximum potential if it only contains a few small to medium-sized trees per hectare (Hitimana *et al.*, 2004). The distribution curve of many Indian forests typically exhibits an exponential decline with increasing GBH (Khamyon *et al.*, 2004), and the same finding is presented in our work.

According to Mani and Parthasarathy (2005), Parthasarathy and Karthikeyan (1997), Srinivas and Parthasarathy (2000), and Willian-Linera (1990), basal area must be taken into account while analyzing the vegetation composition and qualitative characteristics of forests. According to Mani and Parthasarathy (2005), Parthasarathy and Karthikevan (1997), Srinivas and Parthasarathy (2000), and Williams-Linera (1990), basal area is one of the crucial factors affecting the vegetation composition and site quality of forests. The basal area of the tree species differed among the 12 plots used in the current investigation. Badagaon Forest Division's mean stand basal area is assessed to be 20.77 m^2/ha. Panda *et al.* (2013) calculated the mean basal area to be 10.47 m^2/ha based on a quantitative assessment of forest vegetation in a 222-ha sampling area in 10 districts of Odisha located in the Southern Eastern Ghats region. Pragassan and Parthasarathy (2010) measured the mean basal area as 16.9 m^2/ha in their study of the tropical dry evergreen forest of the Eastern Ghats. According to a study of the diversity of tree species in the tropical moist deciduous forest of the Nayagarh forest division, Eastern India, the mean basal area is 16.91 m^2/ha (Sahoo *et al.*, 2017).

The important value index offers a great deal more detail about a species. By looking at a species' IVI value, one can determine whether or not it can survive in a forest area. The variety of tree species present and the level of habitat disturbance both affect IVI value. The IVI value in the current study ranges from 0.77 to 65.47. Shorea robusta received the highest IVI score. The IVI range in Malyagiri's dry deciduous forest is 0.01–44.76, with Shorea robusta recording the highest IVI value (Sahu *et al.*, 2012). For the same species (Shorea robusta), higher IVI values in the Badagaon forest range and somewhat lower IVI values in the Malyagiri hill range suggest that the climatic conditions in Badagaon are more suited for the species than those in the Malyagiri hill range.

According to Sobuj and Rahman (2011) and Deka *et al.* (2012), a low value for species diversity suggests that the ecosystem is not well-developed, whereas a high value for the Shannon-Wiener index shows that the ecosystem is extremely rich. The adaptation of species and the community's stability are both direct determinants of species diversity (Knight, 1975). Generally speaking, the Shannon diversity index (H') is larger for tropical forests; however, the Shannon's index ranges from 0.83 to 4.0 for Indian forests (Singh *et al.,* 1984). Shannon's index variance in the current study ranges from 0.89 to 2.75, which is within the acceptable range. The 11 plots in the research area have an average Shannon diversity index value of 2.05. For the study area, the Simpson and evenness indices are 0.74 and 0.59, respectively. Simpson's index in the current study ranges from 0.35 to 0.86. The diversity and richness of species found in tropical forests around the world vary greatly between dry and humid zones. According to studies, species richness ranges from 20 species per hectare in the flooded Varzea forest of Rio Xingu, Brazil (Campbell *et al.,* 1992) to 307 species per hectare in the Amazonian region of Ecuador (Valencia *et al.,* 1994).

Out of all species in this study, 57 per cent were contiguous, 22 per cent were random, and 4 per cent were regular per hectare, while in a one-hectare region of tropical dry deciduous forest in the Malyagiri hill ranges, Eastern Ghats, India, 38 per cent were contiguous and 4 per cent were randomly distributed (Sahu *et al.,* 2012). According to Odum (1971), the most prevalent type of plant distribution in nature is contiguous distribution. In the foothill forests of the Garhwal Himalayas, according to Kumar and Bhatt (2006), the majority of the species follow a contiguous distribution pattern. Rao and Hajra (1990) discovered a similar finding for tree species in a subtropical forest in northeast India.

When examining the diversity of tree species, each individual species is important. According to Condit *et al.* (1998); Denslow (1995); Hayenk and Buzas (1997); Preston (1962), species richness is a direct cause-and-effect relationship between species abundance. The description of forest composition and the classification of forest types frequently use the most numerous species (Valencia *et al.,* 2004). Only one species, *Shorea robusta*, predominated in the Badagaon forest range of the Similipal, accounting for 33.85 per cent (239 individuals) of the total tree population and 25.56 per cent (5.31 m^2) of the research site's total basal area.

The information on the species-to-genus ratio is highly helpful in establishing the necessity of a certain species, and this value also tells us enough about the species' rate of evolution. According to Mishra (2008), the high percentage suggests recent diversification. The species-to-genus ratio in the study of the current data is 1.07, indicating that species have evolved over a considerable amount of time (Ricklefs and Miller, 2000).

Conclusion

The present study showed the structure and composition of a disturbed forest stand of SBR. The study makes it pretty evident how anthropogenic pressure affects the species richness and density of the forest, but no similar conservation effort was observed for the area. As a result, the vegetation will eventually disappear completely. Thus, this study has significant implications for the preservation of the region's biodiversity. For the purpose of preserving the habitat, integrated conservation is recommended. Both the economic and social well-being of the local population and the conservation of biodiversity will be aided by this. A major contributor to the decline in biodiversity is the over- and under exploitation of natural resources in tropical forests to meet fundamental requirements of the nearby villagers. Therefore, broad awareness campaigns, resource usage that is based on science, and strong administration are urgently needed. The gathering and analysis of long-term ecological data, which calls for scientific baseline research, may be the basis for the SBR forest management problem. Community involvement and the application of conventional technologies can both be used as tools for conservation and ought to be combined to ensure sustainable resource management.

REFERENCES

Ayyappan N and Parthasarathy N (1999) Biodiversity inventory of trees in a large-scale permanent plot of tropical evergreen forests at Varagalaiar, Anamalais, Western Ghats, India. Biodiversity Conservation 8: 1533-15554.

Campbell DG, Stone JL and Rosas A (1992) A comparison of the phytosociology of three floodplain (Varzea) forests of known ages, Rio Jurua, Western Brazilian Amazon. Bot J Linnaean Soc. 108: 213–237.

Census of India (2011) Distrct Census Handbook Mayurbhanj, Village and Town directory, Series-22, Part XII-A. Directorate of Census operation Odisha.

Census of India Report (2001) Office of the Registrar General and Census Commissioner, Ministry of Home Affairs, Government of India, New Delhi.

Census of India Report (2007) Office of the Registrar General and Census Commissioner, Ministry of Home Affairs, Government of India, New Delhi.

Champion HG and Seth SK (1968) A Revised Survey of the Forest Types of India. Manager of Publication, ISBN 81-8158-0613, New Delhi.

Condit R (2000) Spatial patterns in the distribution of tropical tree species. Science 288:1414–1418.

Condit R, Sukumar R, Hubbell SP, Foster RB (2000) Predicting population trends from size distributions: a direct test in a tropical tree community. Am Nat 152: 495–509.

Curtis JT and Cottam G (1962) Plant Ecology Workbook. Burgers Publishing Co., Minnesota: 163.

Curtis JJ and Mclntosh RP (1950) An upland forest continuum in the prairie forest border region of Wisconsin. Ecology 32(3): 476-496.

Dash M and Behera B (2012) Management of Similipal Biosphere Reserve Forest. Advances in forestry letter 1(1): 7-15.

Dash M and Behera B (2013) Biodiversity conservation and local livelihoods: a study on Similipal biosphere reserve in India. Journal of rural development 32(4): 409-426.

Deka J, Tripathi PO and Khan LM (2012) High Dominance of *Shorea robusta* Gaertn. in Alluvial Plain Kamrup Sal Forest of Assam, N.E. India. International Journal of Ecosystems, 2(4): 67-73.

Denslow JS (1995) Disturbance and diversity in tropical rain forests: the density effect. Ecol Appl 5: 962–968.

[FAO] Food and Agriculture Organization of the UN (2010) The Global Forest Resource Assessment, Food and Agricultural Organization, Rome.

[FAO]Food and Agriculture Organization of the UN (2008) Regional initiative for the development and implementation of national level criteria and indicators for the sustainable management of dry forests in Asia.

Gamble JS and Fischer CEC (1915-1935) Flora of Presidency of Madras Vol. 1-3. Adlard and Son Ltd, London, UK.

Gopalkrishna SP, Koong ML, Somashekar RR, Suresh HS and Suresh R (2016) Tree diversity in the tropical dry forest at Bannerghatta National Park in Estern ghats, Southern India.Europian Journal of Ecology 2: 12-27.

Hayek LAC and Buzas MA (1997) Surveying natural populations. Columbia University Ress, New York: 448.

Hegner R (1979) Nichtimmergrüne Waldformationen der Tropen. Untersuchungen zu ihrer Typologie and Verbreitung. Kölner geograph, Köln, Germany: 410.

Hett JM and Loucks OL (1976) Age structure models of balsam fir and eastern hemlock. J Ecol 64:1029–1044

Hitimana J, Kiyiapi JL and Njunge JT (2004) Forest structure characteristics in disturbed and undisturbed sites of Mt. Elgon moist lower montane forest, Western Kenya. For Ecol Manag194: 269–291.

Holdridge LR (1967) Life zone ecology. Tropical Science Centre, San Jose, Costa Rica.

Huston MA (1994) Biological Diversity: The Coexistence of Species in Changing Landscapes. Cambridge University Press, Cambridge.

Indian State of Forest Report (2019) Forest survey of India (vol.I-II) Ministry of environment, forest and climate change,Government of India.

Khamyong S, Lykke AM, Seramethakun D and Barfod AS (2004) Species composition and vegetation structure of an upper montane forest at summit of Mt. Doi Inthanon, Thailand. Nordic J Bot 23: 83–97.

Knight DHA (1975) Phytosociological analysis of species rich tropical forest on Barro Colorado Island, Panama. Ecol Monogr. 45: 259–284.

Kumar M and Bhatt V (2006) Plant biodiversity and conservation of forests in foot hills of Garhwal Himalaya. Journal of Ecology and Application 11(2): 43-59.

Kumar S, Dhankhar S, Arya VD, Yadav S and Yadav JP (2012) Antimicrobial activity of Salvadora Oleoides Decne against some microorganisms. Journal of Medical Plants Research 6(14): 2754-276.

Lamprecht H (1989) Silviculture in the tropics. Eschborn, Rossdorf,Germany: 296.

Mani S and Parthasarathy N (2005) Biodiversity assessment of trees in five inland tropical dry evergreen forests of peninsular India. Syst. Biodiversity 3(1): 1–12.

Mishra BK (2008) Conservation and management effectiveness of Similipal Biosphere Reserve, Orissa, India. The Indian Forester 136(10):1310-1326.

Mishra RC (1997) The status of rare, endangered and endemic flora of Similipal forest: A Natural Habitat of Unique Biodiversity. Environmental Society, Bhubaneswar.

Mohapatra A, Reddy C, Chandrasekhar and Biswal AK (2008) Aquatic plant diversity of SBR, Orisha, India.

Moran D (2018) Tropical deciduous forest, Avian report.

Mori SA, Boom BM, Carvalho AM and Dosantos TS (1983) Southern Bahian moist forests. Bot Rev 49: 155–232.

Murphy PG and Lugo AE (1986) Ecology of tropical dry forest. Annual Review of Ecology and Systematic 17: 67-88.

Odum EP (1971) Fundamentals of Ecology. 3rd eds. W. B. Saunders Co., Philadelphia, USA.

Panda P, Sahoo T and Acharya L (2017) Structure and composition of and diversity of tree species in tropical moist deciduous of Eastern ghat, India: a case study of Nayaghar forest division, Odisha. Journal of Forestry Research: 1-28.

Panda PC, Mahapatra AK, Acharya PK and Debata AK (2013) Plant diversity in tropical deciduous forests of Eastern Ghats, India: A landscape assessment. International Journal of Biodiversity Conservation 5(10): 625-639.

Parthasarathy N (1999) Tree diversity and distribution in undisturbed and human-impacted sites of tropical wet evergreen forests of Southern Western Ghats, India. Biodiversity Conservation 8: 1365-1381.

Parthasarathy N and Karthikeyan R (1997) Plant biodiversity inventory and conservation of two tropical dry evergreen forests on the Coromandel Coast, South India. Biodiversity Conservation 6: 1063–1083.

Parthasarathy N and Sethi P (1997) Tree and liana species diversity and population structure in a tropical dry evergreen forest in South India. Trop Ecol 38: 19-30.

Poorter L, Bongers F, Rompaey RSAR and Klerk M (1996) Regeneration of canopy tree species at five sites in West African moist forest. Ecol Manag 84: 61–69.

Pragasan A and Parthasarathy N (2010) Landscape-level tree diversity assessment in tropical forests of southern Eastern Ghats, India. Flora 205(11): 728–737.

Pragassan A and Parthasarathy N (2010) Landscape- level tree diversity assessment in tropical forests of Southern Eastern Ghats, India. Flora 205(11): 728-737.

Preston FW (1962) The canonical distribution of commonness and rarity. Ecology 43: 185–215.

Ramakrishna A, Siddiqui SZ, Sethi P and Dash S (2008) Founal Resources of Similipal Biosphere Reserve, Mayurbhanj, Orissa., Conservation area series 28:1-87.

Rao RR and Hajra PK (1990) Distribution of vegetation types in northwest Himalaya with brief remarks on Phytogeography and Floral Resources Conservation. Proc. Indian Acad. Sci (Plant sci.) 100(4): 263-277.

Reddy CS, Sahu SC, Dhal NK, Pattanaik C and Brahmam M (2007) Phytosociological study of Tropical dry deciduous forest of Boudh District, Orissa, India. Research journal of Forestry 1(2):66-72.

Ricklefs RE and Miller GL (2000) Ecology- W.H. Freeman and Company, New York.

Rout SD (2008) Anthropogenic Threats and Biodiversity Conservation in Similipal Biosphere Reserve, Orissa, India. Tiger Paper 35 (3): 22 -26.

Roy PS, Dutta CBS and Joshi PK (2002) Tropical Forest Resource Assessment and Monitoring. Tropical Ecology 43(1):21-37.

Sahu SC, Dhal NK and Mohanty RC (2012) Tree species diversity, distribution and population structure in a tropical dry deciduous forest of Malyagiri hill ranges, Eastern ghats, India. International society for tropical ecology 53 (2): 163-168.

Sahu SC, Dhal NK, Reddy CS, Pattanaik C and Brahmam M (2007) Phytosociological study of tropical dry deciduous forests of Boudh District, Odisha, India. Res J For 1(2): 66-72.

Saxena HO and Brahmam M (1994-96) The Flora of Orissa, Vol I-IV. Orissa Forest Development Corporation Limited, Bhubaneswar.

Saxena HO and Brahamam M (1989) The flora of similipahar (similipal), Orisha. Regional Laboratory (CSIR), Bhubaneswar.

Saxena HO and Brahmam M (1989) The flora of Similipahar (Similipal), Orissa with particular references to the potential economic plants. Regional Reseach Laboratoty, (CSIR), Bhubaneswar: 231.

Shannon CE and Weaver W (1963) The mathematical theory of communication. Urbana, IL; University of Illinois Press.

Simpson EH (1949) Measurement of diversity. Nature. 163(4148): 688.

Singh JS and Chaturvedi RK (2018) Tropical dry deciduous forest: Research trends and emerging feature. Library of Congress: 1-248.

Singh JS, Singh SP, Saxena AK and Rawat YS (1984) India's Silent valley and its threatened rain forest ecosystem. Environ Conserv. 11: 223–233.

Sobuj NA and Rahman M (2011) Assessment of plant diversity in Khadim nagar National Park of Bangladesh. International Journal of Environmental Sciences 2(1): 1-13.

Srinivas V and Parthasarathy N (2000) Comparative analysis of tree diversity and dispersion in the tropical lowland evergreen forest of Agumbe, Central Western Ghats, India. Trop Biodivers7: 45–60.

Sukumar, R, Dattaraja, HS, Suresh, J, Radhakrishan, R, Vasudeva, S, Joshi, NV (1992) Long term monitoring of vegetation in a tropical forest of Mudumalai, Southern India. Curr. Sci. 62: 608-616.

Thomas SC and Baltzer JL (2002) Tropical forests. United Nations Decades on Biodiversity: 1-8.

Valencia R, Balslev H and Mino GCPY (1994) High tree alpha-diversity in Amazonian Ecuador. Biodivers Conserv 3: 21–28.

Valencia R, Foster RB, Villa G, Condit R, Svenning JC, Herna´ndez C, Romoleroux K, Losos E, Magard E and Balslev H (2004) Tree species distributions and local habitat variation in the Amazon: large forest plot in eastern Ecuador. J Ecol 92: 214–229.

Whitford PB (1949) Distribution of woodland plants in relation to succession and clonal growth. Ecology 36: 117-127.

Whitmore TC (1998) An Introduction to Tropical Rain Forests, Second ed., Oxford University Press, New York.

Williams- Linera G (1990) Vegetation structure and environmental conditions of forest edges in Panama. J Ecol 78: 356–373.

Wilson EO (1988) The current state of biological diversity. Biodiversity: 3-1.

Chapter 17

Ecology, Traditional Knowledge and Economic Importance of Bamboo in Mayurbhanj District of Odisha

Sifan Priyadarshini, Shakti Swarupa Barik and Sudam Charan Sahu

Department of Botany, Maharaja Sriram Chandra Bhanja Deo University, Sriram Chandra Vihar, Takatpur, Baripada–757003, Odisha
e-mail: sudamsahu.bdk@gmail.com

Introduction

The connection between man and nature dates from the far past. Trees are one of the greatest blessings for humans from mother Earth. Trees make our environment beautiful and intimate. They help in making a healthy community life. Along with their countless benefits they are the best friends of human beings since pre-historic time and use of plants or plant products is as old as mankind itself. Without trees, life seems impossible. Most crucial components for survival such as oxygen and food are derived from plants only. The protective layer which guards us against all kinds of damages and harms are only provided by nature. As we evolved, we started depending on trees for medicine, shelter and other commercial uses. Every form of nature; the plants, animals, mountains, moon, rivers and more significantly important for our lives. Absence of one element is enough to cause a catastrophe in the functioning of human life. Even today, our dependence on trees has not reduced but increased than ever before.

Forest is one of the precious resources provided by nature to mankind that just keeps on giving. Its existence has taken care of mankind and nourished it long before humans. The significance and usefulness of forests date back to those days when man learnt the merits of a settled life. So, the survival of mankind without nature is impossible. Forest serves people in many tangible ways. They are an inexhaustible reserve and provide man and industries with numerous raw materials. They constitute a perennial economy, unlike minerals that, once mined, cannot be renewed. There are many forest products which are the sources of livelihood for the rural areas people.

In this modern world, forest product like Bamboo, occupied an important place in economy of any country and betterment of forest dependent poor people. Bamboo is a composite natural wood whose strength is generally similar to a strong softwood or hardwood. Its strength-to-weight ratio is similar to timber and mostly used in construction works. In recent years, bamboo has been served as the most economic important plant, textile industries have also demands on Bamboo clothing as is produces antibacterial and UV absorbant, due to presence of lignin in in bamboo fibre. Bamboos, tallest grass of the world belongs to the family Poaceae and sub family Bambusoidaceae. Unquestionably it is the most important plant family, supplies itself as food for both herbivores and mankind. It covers a wide range of needs of human being and protect environment. Bamboo grows in tropical, sub-tropical and mild temperate zones of the globe. These are tremendously diverse plants, which have the capacity to adapt any extreme climatic and soil conditions. Most of the bamboos are found in forest and also widely spread outside the forest, usually in the farm lands, river banks, road sides and rural areas.

Long after the demise of the dinosaurs, grasses are relatively the recent additions to the flora of globe. About 1250 species under 75 genera of Bamboo are found growing all over the world. In India, bamboo is represented by 136 species under 20 genera. India accounts for more than 50 percent of the total bamboo reserves in the world. Bamboo occupies 8.9 million hectares of forest area in India, which accounts for a little more than one eighth of the total forest area of the country. More than 50 percent of the bamboo species occur in Eastern India.

About 13 species of bamboo are present in Odisha (Dash *et al.,* 2020). The other areas, which have a significant presence in the bamboo map of India, are the Andamans, Bastar region and the Western Ghats. The total bamboo forests in the country annually produce 4.5 million tonnes of bamboo. Though found in almost all parts of Odisha in all types of lands including commons and revenue land. Bamboo is found more in the forests of Central and Southern Odisha. Even coastal Odisha had relatively fair availability of bamboo till the cyclone of 1999. In the dry deciduous forestsof the state, major species composition is Sal, Teak and Bamboo. In the moist deciduous forests of Odisha, Sal is the predominant species followed by Bamboo. There are 374.77 sq.km. of pure bamboo forests and 177794.61 sq.km.

of mixed bamboo forests in the state. In Odisha among the 13 species of bamboo, 4 to 5 species are common. These include *Dendrocalamus strictus* (Roxb.) Nees (Salia baunsha), *Bambusa bambos* (L.) Voss (Kanta baunsha), *Bambusa nutans* Wall. ex Munro (Badiabaunsha) and *Bambusa tulda* Roxb. (Sunderkani/Golden bamboo). Among these wild species, *Dendrocalamus strictus* and *Bambusa bamboos* are cultivated. Whereas *B.nutans* and *B.tulda* are mostly found growing in village ares rather than wild zones. *Dendrocalamus strictus* grows suitably in all kind of soil and is known as the male bamboo or hill bamboo. While *Bambusa bambos* prefers moist and rich soil for its growth. It is generally found growing abundantly on-stream banks. Thorns make up their clump impenetrable. Due to the longest and wider culms *Bambusa nutans* is much preferred by growers. It may be noted that *B. nutans* is scarily distinguishable from *B.tulda* and is comparatively thicker than *Dendrocalamus strictus.* It is comparatively soft to work with *D. strictus* as it is less prone to insect attack when proporly dried. It is quite difficult to work with *B. bambos* due to the presence of thorns.

Bamboo, strong as strongest wood is available at the much lower price as compared to wood. So, it is called as the "Green gold" of the twenty-first century. It is a replacement of the depleting and costly wood resources. East Asia and South America are considered to be the centre of diversity of bamboo. Bamboos cover more than 18 million hectares of land, of which 75 percent is in Asia. About 2.5 billion People worldwide use bamboo and 1 billion people live in bamboo houses. Bamboo grows three times as first and can be harvested four times as often as *Eucalyptus.*

The life span of bamboo is only about 20 years and it flowers once in its life cycle. After growing from rhizomes and generating a number of culms for a period of 50-80 years, nearly all the clumps of bamboo lose many culms as they produce wind-pollinated flowers, set free large quantities of seeds and die synchronously. Regardless of their geographical location, a particular species will flower at a time. So, the environmental factors does not impact on flowering of the species regardless it indicates the presence of some sort of "alarm clock" in each cell of the plant. Each cell transmits all energy for flower production that indicates end of vegetative phase.

Bamboo culms emerge individually from the ground and grow to the full diameter independently. Within three to four months, they attain their full height. At this time the nodes give rise to new branches and leaves starts arising. The pulpy wall of each culm slowly hardens and the shoot becomes fully mature within three to four years. Individual species produces individual small culms initially and they mature to attain a particular height and harvested. Bamboo is very close to the local community of villages as it provides the basis for fulfillment of every day necessity to the community. People use bamboo according to their traditional knowledge and customs concerning the plant. Despite of different climatic conditions bamboo grows and flowers at the same time. Moreover, it is versatile and incredibly flexible; it bends in strong wind rather than breaking.

This study is mainly focused on studying the diversity, traditional knowledge of bamboo in Mayurbhanj district. It also aims to cover the economic importance of this amazing plant for the local people of different areas of this district and what measures can be taken so that bamboo can be used to the fullest by which the people can get more benefit and can conserve their environment. This study also gives a message about the eco-friendly nature of the bamboo products which can be utilized by the people instead of the plastic or other non-biodegradable materials which cause harm to the environment.

Study Area

Mayurbhanj, the largest district of Odisha state is a part of Eastern India. It covers about 10,418 sq.km and it lies between 21°16' and 22°34'N and 85°40' and 87°11'E. It is one of the tribal dominated districts of the state. Agriculture occupies a vital role in the economic development of the district and many people depend on forest for their livelihood. The forest products have a great role in the socio-economic life of the tribal people in fact their life can't be imagined without forest. The tribal people collect various forest products to satisfy the needs for their food, fodder, medicine and other uses. These people have a major role in the conservation of forest and no one can look after their forest and environment better than them as their survival and identity depend on it. In other words, district is home to the Similipal Biosphere Reserve. So the tribal people and the forests have an inter-relationship with each other where both get benefitted positively.

Bamboos have been traditionally used by the people of the district in many ways. The bamboos usually found in forest areas and also near the road sides, river banks. This plant contributes significantly to the tribal community's economic benefits in every day, where production and process of the goods made up of bamboo significantly supports basic needs and livelihood earning. The exquisite creations are all made lovingly by skilled women weavers, who deftly transform the ordinary raw materials into stunning and functional works of art one can take back home as unique souvenirs. Apart from the tribal people bamboo is used by many villagers for various purposes. The study was conducted in different areas of Mayurbhanj specifically in four blocks-Gopabandhunagar, Khunta, Udala and Kaptipada (Figure 17.1).

Soil

Matured Red Lateritic soil, mixed grey soil and Unaltered soils with coarse parent materials are the different soil types of the district. Red-laterite soil covers most parts of the district which includes Bamanghati and Panchpir plateau.

Climate

Similipal being the home to the district, it influences the weather substantially. A sub-tropical climate with a hot summer, chilling winter with good precipitation is experienced in Mayurbhanj. The average annual rainfall is around 164 cm. During

Summer temperature reaches upto 45°C occasionally but thunderstorms in the evening show moderating influence. Minimum temperature in winter goes down to 8 degrees.

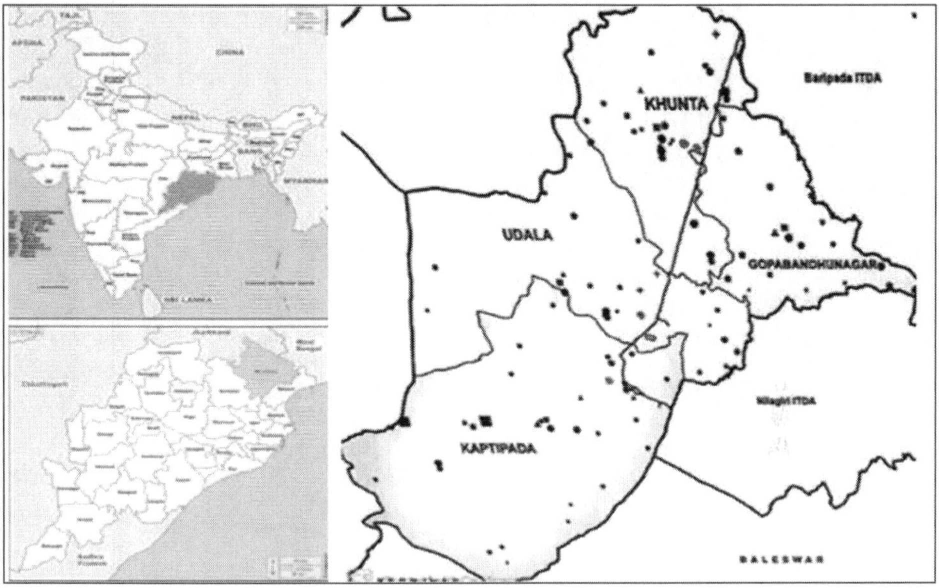

Figure 17.1: Map of Study Area.

Vegetation

Lush green vegetation is the uniqueness of the district. It has mainly dry deciduous and mixed deciduous forest. The forests are mainly dominated by plants like Sal, Mahula, Asana and other indigenous plant species which provide many benefits to the local people. The bamboos are found throughout the district in the forests as well as in the village lands which have great utility for the people of the district.

Methodology

Data Collection

Mayurbhanj being a tribal dominated district, the tribes constitute about 58.7 per cent of the total population as per the 2011 census. The field study on bamboos of Mayurbhanj district was carried out during 2020-21. Data were collected from previous literatures and interviewing the villagers of different blocks about the uses and traditional knowledge of bamboo. The information on types of bamboo growing in this district (local names of bamboo), types of bamboo used by the community, where the bamboo taken, usefulness of bamboo in everyday life, parts of the plant used, how to use were collected through direct observation supported by semi structured questionnaires, personal and group interviews of local communities especially bamboo craftsmen of the district and market surveys.

There are weekly village markets (locally known as 'Hata') from which much information were collected about the bamboo works which are the source of income for many villagers. Photographs of different species of bamboo (Figure 17.1) and the materials made from it were taken and represented in (Figure 17.3). During these interaction with the tribal community of the district it is realized that how important are the plants for human beings and to what extent the plants can help us. All the species of bamboo studied were mentioned with their botanical names as well as their local names. Identification of the bamboo species was done with the help of Flora of Orissa (Saxena and Brahmam, 1994-96) and other available literatures.

Soil Collection and Analysis

Soil samples from different bamboo dominated areas of Mayurbhanj district were collected in zip lock collection bags up to a depth of 30 cm. Then these soil samples were brought to laboratory for processing and analysis.

Results

Soil Analysis

Soil pH is a measure of the acidity or alkalinity of a soil. It is a key characteristic that can be used to make informative analysis both qualitative and quantitatively regarding soil characteristics. From analysis of these soil samples it is found that the pH of the soil ranges from 4.5 to 4.6 indicating the acidic nature of the soil. Soil Electrical Conductivity indicates the soil health that measures salinity of soil. Here, the Electrical Conductivity (EC) value ranges from 0.02 to 0.04 ds/m.

Table 17.1: Soil Nutrient Analysis in Four Sites of Mayurbhanj (Results are in Mean ± S.D).

Site No.	*pH*	*EC (Electric Conductivity) (ds/m)*	*OC (Organic Carbon) (per cent)*	*Available Phosphorous (P)(kg/ha)*	*Available Potassium (K) (kg/ha)*
Site-1	4.53±0.05	0.03 ±0.01	0.88± 0.39	12.0 ±18.65	366.33± 107.05
Site-2	4.56±0.01	0.04± 0.01	0.86 ±0.39	5.43 ±1.96	337.33 ±124.04
Site-3	4.60±0.17	0.02 ±0.01	0.93± 0.30	5.40 ±2.00	378.33 ±105.15
Site-4	4.52±0.04	0.03±0.02	0.81±0.35	5.30±1.86	365.33±104.10

Phosphorus' primary role in a plant is to store and transfer energy produced by photosynthesis for use in growth and reproductive processes. Here the observed value of phosphorus ranges between 3.60 kg/ha to 27.2 kg/ha. Potassium is absorbed by plant from the soil that increases disease resistance, helps stalks to grow upright and sturdy, improves drought tolerance and helps plants get through the winter. These collected soil samples have potassium values ranging from 255.00kg/ha to 485.00kg/ha. Higher organic carbon promotes improves soil aeration (oxygen in the soil) and water drainage and retention and reduces the risk of erosion and nutrient leaching. The soil organic carbon value of these collected soil samples ranges from 0.47 per cent to 1.25 per cent.

Taxonomy of Bamboo

Ecologically bamboos are found in the forest floor, river banks or planted in gardens. From the field observation, it was found that local communities do bamboo plantation for commercial purposes that includes *B. bambos*, *B. tulda* and *B. nutans* and *Dendrocalamus strictus.*

Morphological Description and Distribution

Bambusa bambos

These are spiny or thorny bamboos, shrubs found in mainly stream banks, deciduous forests and also in home gardens. Flowering of this species occurs once in 100 years. This species can be identified by large thorny appearance with erect and woody bright green culms.

Bambusa tulda

These are very large, spineless bamboos up to 21m high and 5-10 cm diameter, white rings below the nodes having hairy or white powder on culms. The upper surface of the sheath is covered with blackish-brown hairs. The young shoots are yellowish green in colour, culm walls are very thick and nodes are prominent. These species are also found in deciduous forests and also inroad sides in the village areas.

Bambusa nutans

These bamboos have erect culms, densely covered with black deciduous bristles at the back and are solitary from a creeping rhizome. *B. nutans* are found in village lands as well in forest areas.

Dendrocalamus strictus

These are arborescent bamboos, whose culms are usually erect from a densly branched rootstock. They have the culm sheaths often very large,elongated and sometimes the leaves are very broad,found in deciduous forests.

Key to the Genera of Bamboos

1. Arborescent, shrubby or rarely climbing bamboos, sometimes thorny, lodicules 3 ....***Bambusa***
1. Arborescent, always unarmed, lodicules absent .....***Dendrocalamus***

Key to the Species of *Bambusa*

1. Branches spinescent ....***B. arundinacea***

1. Branches un-armed
    2. Caespitose, palea not ciliate, anthers obtuse .... ***B. tulda***
    2. Culms solitary, palea long-ciliate, anthers apiculate ....***B. nutans***

1. B. bambos

Large thorny bamboo, branches from the base, caespitose. Culms yellowish-hairy when young otherwise glabrous. Leaves linear-lanceolate, leaf-sheath striate, blade triangular, sharply pointed. Ligule short. Palea ciliate. Glumes many nerved. Spikelets sessile in close dense clusters. Anthers obtuse.

Figure 17.1: Diversity of Bamboos in Mayurbhanj District of Odisha (A. *Bambusa tulda*, B. *Bambusa nutans*, C. *Bambusa bamboos*, D. *Dendrocalamus strictus*).

2. B. tulda

Very large bamboo upto 21m high and 5-10 cm diameter, caespitose, white ringed below the nodes, culms having hairs or white powder. Leaves linear-oblong, base usually rounded, petiole often hairy, leaf-sheath glabrous, spikelets in interrupted clusters, Palea shorter, anthers obtuse, glabrous.

3. *B.nutans*

Culms solitary from a creeping rhizome, culm-sheaths 15-30 cm, densely covered with black deciduous bristles at back, blade triangular- acuminate, leaves lanceolate-linear, ligule long,obtuse, hairy. Spikelets in bracteates fascicles.Glumes 2-3, lemmas 3-7 fertile. Palea shorter,ovate, flattened. Anther-tips long apiculate or slightly penicillate. Ovary subovoid.

Table 17.2: List of Bamboo Species with their Local Names and Chemical Constituents Found in Mayurbhanj District of Odisha

Scientific Name	*Local Name*	*Distribution*	*Chemical Constituents of Leaves (Sahoo et al., 2010)*			
			C	*CHO*	*CP*	*HC*
Bambusa tulda	Sunderkani/ Golden bamboo	Desiduous forest and road sides.	382	702	146	317
Bambusa nutans	Badia baunsha	Desiduous forests, village lands.	366	675	182	346
Bambusa bambos	Kanta baunsha	Stream banks, village gardensand deciduous forests	374	645	184	341
Dendrocalamus strictus	Salia baunsha	Desiduous forests	446	678	161	243

Note: C- Cellulose, CHO- Total Carbohydrate, CP- Crude protein, HC- Hemicellulose.

Traditional Knowledge and Uses

A strong relationship between people and bamboo is observed in Mayurbhanj district. Bamboo plays a great role in many cultural, social, and economic factors including beliefs, customs and habits. A number of traditional uses and practices of bamboo observed in Mayurbhanj district are described and demonstrated in this paper with photographs (Figure 17.3). Of these, uses for construction materials, handicrafts, furnitures, traditional ceremonies, transportation and food are noteworthy.

Construction and Materials

The culm characteristics usually determine the utilization of bamboo. In the areas of construction and infrastructure it plays the major role. It is used structurally in buildings for making home materials such as walls, roofs, doors, windows as well as poles. Straight, thick, and sturdy bamboo culms are mainly used for making the main pillars of a building. For the construction of mud houses, first the walls of the house are made from the spiny branches of bamboo and then mud is daubed on to the walls on both sides (Figure 17.3: 2). On the other hand, the local people of village and forest areas, where they have no facility of concrete bridge use bamboo to make bridge to cross river.

Bamboos are also used during the construction of concrete buildings. After the construction of roofs bamboos are used as pillars for reinforcement of the newly made roofs and after the roofs become strong the bamboos are removed. Bamboo scaffoldings are popular for use in high-rise buildings. To protect own house area or vegetable farms, people of village are make bamboo fences in which they mainly use the spiny branches of bamboo so that no grazing animal can enter into the farm to eat the plants and the fences of houses also help to keep away outsider animals from the house (Figure 17.3: 6).

Handicrafts and Furniture

From the study it is found that there are many people in this district whose life revolve around bamboo. Mainly there is a community of people, known as "Dom" community which is a scheduled caste tribe. The people of this community are closely associated with bamboo and they are almost completely reliant on their occupation of bamboo works. They make different types of baskets, brooms and hand fans using bamboo. Baskets made up of bamboo are used for various purposes, some baskets locally called as "Pachhia" are used for storing grains; large baskets called "Jhampi" are used to keep hens (Figure 17.3: 17). Other small bamboo baskets are used in house hold purposes like keeping vegetables or kitchen items. Some wide bamboo baskets are made of bamboo called as "Charni" by the tribals which are used to dry the boiled rice used for making the famous tribal drink Handia (Figure 17.3: 18).

The bamboo craftsmen have their special weaving skills by which they make all these materials. After weaving they also colour these materials beautifully to make them attractive for the buyers. These bamboo products have very good demand in the village weekly markets. When one will visit these markets he will find a nice collection of bamboo crafts which are brought by the bamboo workers and nicely decorated by placing the baskets and other bamboo materials one over another. People also make the frame of bed using bamboo which is locally called "khata" and they complete the bed by weaving with thin ropes (Figure 17.3:15). Other materials like ladder, fish trap and mats are made using bamboo in the village area. Special types of bamboo containers are made which are commonly called as "Khalei"for the storage of both fresh as well as dried fishes (Sukhua) (Figure 17.3: 23). Using bamboo large storage baskets called "Duli" are also made which are specially used for storing paddy for long time (Figure 17.3: 30). But now-a-days these are rarely used as the plastic bags are used more for this storage purpose.

Figure 17.3
1-2: Bamboo house frames; 3-4: Bamboo doors;

Figure 17.3

5: Bamboo poles used in concrete house; 6: Bamboo fence; 7: Gate made up of bamboo; 8-11: Different types of bamboo baskets.

Figure 17.3

12: Winnowing basket of bamboo; 13: A women weaving bamboo basket; 14: Bamboo fish trap; 15: Bamboo frame of bed; 16: Paddy threshing instrument and ladder; 17: Large bamboo basket used to keep hens.

Ananda Mohanta, a well-known renowned bamboo artisan of Mayurbhanj district is famous for making different bamboo furnitures like swings, sofa-sets, cottage-style houses, doors, windows and 40 different products. He has been formally trained in bamboo craft at the National Bamboo Mission in Dehradun. He makes several other household items starting from bowls, pen stands and water bottles. Bamboo is also used to make many other materials like handles of umbrella and farming tools like spade, axe and sickle *etc.*. Walking sticks are also made from bamboos which mainly help the old persons as a support for walking.

Figure 17.3

18: Bamboo basket to dry boiled rice to make the tribal drink called Handia; 19: Bamboo baskets worshipped during Makarsankranti; 20: Bamboo swing; 21-22: Bamboo baskets used in Santali marriage; 23: Fish kept in a bamboo container; 24: bamboo mat; 25: A boy catching fish by bamboo fishing rod.

Figure 17.3

26: Bamboo cage; 27: Edible bamboo shoots; 28: Bamboo as fuel; 29: Another type of fish trap; 30: Fish keeping container; 31: Bamboo trellis; 32: Bamboo broom.

Festivals and Traditional Ceremonies

The proverb saying "Bara masa Tera parba" of Odisha *i.e,* 13 festivals in 12 months are also celebrated in Mayurbhanj district of Odisha. All of these festivals have certain specialities. In some of the festival's bamboo is used in different ways by the people. Makar sankranti is an important festival of Odisha as well as in Mayurbhanj celebrated in the month of January (Odia month Magha). The festival marks the transition of the Sun into Capricon on its celestial path. This festival is observed in this district largely by the people with great joy, especially by the tribals since it falls just after the harvesting of the paddy crops is over. On this day of Makarsankranti women of most of the houses prepare a special type of bhog called Makara Chaula prepared with newly harvested rice, jaggery and fruits and offer it to sun. At the place of worship they place bamboo baskets off large to small sizes one on another successively where the largest basket lies at the bottom, while the smallest one is at the top, as a ritual and worship these baskets as these baskets are used during harvesting crops. The smallest bamboo basket is called 'Katha' which is mainly used to measure rice and paddy. In the next day the children of the family have to push these baskets to make them fall from early morning. Another interesting thing is that people in the early morning *i.e.* before the sun rise of Makar Sankranti set fire to a large bamboo culm which is cut and placed in the middle of a field along with some dry branches like branch of date palm tree which is called 'Makarapala'. There is a belief that this ritual is performed to wardoff winter as during the month of January Mayurbhanj experiences a chilling winter due to the hills of Similipal.

Raja parba is another big festival celebrated in this district for three days in which swings are made with ropes or bamboo in villages where the girls and children celebrate this great festival with joy and merriment by swinging. But bamboo swings have some speciality that many children can swing at a time in these swings and these are very strong. The tribal people have dominant population in this district and they have many interesting traditional ceremonies which they perform with great amusement. One of the tribal communities is known as Santal community. Marriage in Santal community is a vital event and is an important milestone in the lifecycle of a Santal individual. Because of it's vital nature, a series of social ceremonies and rituals are associated with it. There are different types of marriages in their community.

During the marriage ceremony a few members of the groom's party put the bride in a bamboo basket(known as 'Daura' in Santal language) and bring her out from the house, then the groom is lifted over the shoulder of a person and both the bride and bridegroom perform the rituals of marriage. In another type of tribe known as "Bhumij" people use bow and arrow which is made up of bamboo during their marriage ceremony. They shoot the arrows while going to bring the holy water which is used to bath the bride and bridegroom. Bamboo is also used in funeral ceremony. When someone dies people make a bamboo bier to take the dead body to the cremation ground which is locally called "Kokei" is made by using six pieces of bamboo.

Food

Dates back, bamboo shoots has been used as food for both herbivores and human beings. These shoots have a high content of protein (amino acids), carbohydrate, minerals, and several vitamins. The people usually process the young shoots of bamboo (locally called as 'Baunsha Karadi') into a vegetable. First the sheaths are removed from the surface of the young bamboo shoots, then the young white shoots are collected, washed, thinly sliced and boiled in boiling water. The bamboo shoot curry or pan cake is prepared and served.

Other Uses

Except the above specific uses bamboos are also used in many other purposes. The bamboo culms are used as tent poles, the thin branches of bamboo are used as fishing rods which are very amusing materials for the village children who sit near the crop fields having water or near the ponds, rivers and catch fish all through the day using the fishing rods.

Dried bamboo culms are also used as fuel by the villagers and the thin bamboo sticks are specially made for making puffed rice (Mudhi). Parboiled or dried pre-cooked rice is added to the heated sand or soil in the pan and stirred using these bamboo sticks then puffing starts almost immediately. Musical instrument like flute is also made from bamboo. Bamboo sticks are also used to handle the cattle during driving bullock cart. People also make bamboo trellises in their vegetable gardens at home to support climbing plants like bitter gourd, bottle gourd, ridge gourd and ivy gourd *etc.*. Dried bamboo leaves are also used for organic fertilizers. In village people usually cook food in mud hearth and use Bamboo as a common fuel.

Ecological Value of Bamboo

Bamboo serves multiple ecological functions. It helps in water conservation and has a wide spread root system and large canopy that works as a natural water control barrier which prevents massive soil erosion. So bamboos are also planted by the people in the bank of ponds without pesticides or chemical fertilizers hence it does not contribute to soil pollution and proper harvesting prevents the loss of the soil of the banks. Generally, bamboo plants produce much more litter than other vegetation, so it is very useful in improving soil hydrological properties. It has comparatively dense foliage which protects against beating rains. The origination of new culms from underground rhizomes makes it safe harvesting without disturbing the soil.

Bamboos also help in controlling the rate of environmental pollutions. It reduces up to 35 per cent carbon dioxide gases blamed for global warming. It helps to create more clean air in the environment *i.e.* it generates 30 per cent more oxygen than other trees. Naturally, Bamboo has a waxy surface which makes it free from health hazards and also reduces the use of timber consumption. Due to the high bamboo biomass stocks and carbon storage, Bamboo plants, plays most important role in improving climate change. Some bamboo sequesters up to 12 tons of CO_2/ha, which makes it an efficient replenisher of fresh air. So it is one of the sources which contribute to the clean and fresh atmosphere of the Mayurbhanj district. Due to its high nitrogen consumption, excess nutrient uptake of waste water from manufacturing, livestock farming and sewage treatment, Bamboo acts as an excellent scavenger; it helps in reducing water pollution.

People of Mayurbhanj use house hold products made from bamboo which are strong and ecofriendly. The diverse of application of bamboo make bamboo environment friendly and easily adaptable. People use this plant from far past times, when technology was not developed. So, from the ecological point of view bamboo is a very useful plant. Bamboos in this district are mainly found in village lands, in the road sides and home gardens also. During monsoon, the clumps slow the flow and reduce chances of flooding downstream. It is a pioneering plant and is grown by people to restore the lands damaged by overgrazing and poor agriculture. Thus, Bamboo act as the most important plant that helps in restoring previously degraded soils back into productive area.

Bamboo being the fastest growing productive plant, it protects and preserves the local ecosystem, regulates water flow, recharges the water table and conserves the flora and fauna of Mayurbhanj district. As well as it provides environmental security for the local people of Mayurbhanj. During the study some associate species of bamboo are also recoded. The associate species of bamboo mainly includes *Shorea robusta* and *Tectona grandis*. Bamboos growing in village lands or home gardens have usually associated plants like *Vitex negundo, Pongamia pinnata, Azadirachta indica*. It is also the shelter of different types of birds like dove, owl, myna, white-breasted waterhen and many reptiles. Many animals like cattle also take rest under the shade of large bamboo clumps in village areas.

Discussion

Odisha holds 13 species of Bamboo (Dash *et al.,* 2020). This study found 4 species of Bamboo growing in Mayurbhanj. Out of those 4 species, the genera *Bambusa* and *Dendrocalamus* are primarily found under tropical conditions. Dry bamboo brakes (mainly *Dendrocalamus strictus* or salia bamboo) are predominant in the dry deciduous forests of Angul, Athamallik, Baliguda, Ghumusur North, Kalahandi, Rayagada, Bamra, Bonai, Rairhakhol, and Sundargarh divisions. Whereas, *Bambusa bambos* that flourishes best in moist deciduous forests is found in Dhenkanal, Nayagarh and many other divisions.

This study reveals that *B. tulda* is predominant in Mayurbhanj district. Bamboo, being an universal used plant material, it contributes in overall development of a region *i.e.,* social, economic and ecological. In India, Bamboo brings as important component of homesteads, unproductive lands and is of common occurrence in the natural forests fulfilling the subsistence needs of more than 2.5 billion (Anonymous, 1994, INBAR, 1999). It is an important component in developing economy of bamboo-dependent sectors of the population and tribal forest dwellers, bamboo craftsman, artisans and local rural people (Nair et.al, 1982; Muraleedharan *et al.,* 2007; Jayasankar, 1996, 2004; Anitha, 2008, 2012). It is estimated that nearly about 8 million people can get employment working on Bamboo sectors (Anonymous, 1994). About 2 million traditional artisans depends almost entirely on the harvesting, processing and selling of bamboo; bamboo products such as baskets, mats and handicrafts as their earnings for survival. From the study it is found that the people of the village area of Mayurbhanj depend on bamboo for their livelihood to a great extent. Some degraded lands become productive due to the plantation of bamboo. Starting from household to many domestic and agricultural uses, ornamentals to houses, it plays an efficient role (Muraleedharan *et al.,* 2001).

The young edible shoots of the bamboo plant are usually harvested before two weeks from initiation when it attains a height of about 20-30cm. The shoots usually emerge after the rainy season. Bamboo shoots are processed before use. Bamboo shoots have high nutritional value and low fat, and are a good source of fiber and rich in vitamins, cellulose and amino acids (Pradhan, 2015). This may be one of the causes for the healthy body of the tribal people of this area. At harvesting, a shoot may contain as much as 90 per cent water and edible content of a newly harvested shoot is typically around 30 per cent as cited by some informants. This study reveals that bamboo is used by the local people of Mayurbhanj largely, but proper knowledge of its plantation and harvesting can bring them more benefit. It is observed that bamboos only grow in the road side or fields naturally but there is less bamboo plantation. There are many degraded lands where no agricultural work takes place. So if the degraded lands are planted with bamboos this can restore the land as well as it can be a great source of economy for the poor village people.

Conclusion

Bamboo, an amazing plant with multiple use values and uncountable benefits for the environment. It plays a significant role in increasing economic value. From this study it is found that people of Mayurbhanj district are more depended on bamboo as compared to other plants. Generally, they depend on specific plants for specific purposes but bamboo is a plant on which they rely to make daily necessary materials for survival. Depending on the height of the plant, whole plant parts are processed into various products for meeting community need that become a source of income of rural livelihood. During the survey it is seen that mostly women are engaged in bamboo crafts making. Hollow chambers of bamboo's stem make it lighter but strong as wood. Women could easily get opportunity in the field of bamboo economy; light weight of bamboo makes them independent worker. As it reduces investment burden for farmers and does not include any loss in bamboo plantation, it becomes a great source of benefit for these simple living peoples. Along with this, Bamboo leaves are generally the food of herbivores.

Although this plant is not cared like other plants which need proper irrigation and application of fertilizers for its growth, yet it provides easy livelihood fulfilling the necessities of human beings as well keeps surrounding less pollution free. Bamboo not only helps in maintaining livelihood of the people but also has a great connection with their social and cultural life. Being a versatile plant bamboo has played a great role in every aspect of human survival from food, ecological perspectives to economic values, Bamboo has become a part of life of rural livelihood. Bamboos re-germinate themselves, there is no need of proper planting of a bamboo if once it's planted in areas of frequent soil erosion or places with less bamboo species of this district. Besides these uses, bamboo is the home to most faunal species. So, there is a need to conserve bamboo so as to conserve the species that rely on it. Additionally, this provides employment opportunities for the rural and tribal communities through the State governments' missions on suitable management and proper utilization of Bamboo in the state.

REFERENCES

Alam MK (2001) Bamboos of Bangladesh: A field identification manual. Government of the people's Republic of Bangladesh, Bangladesh Forest Research Institute, Chittagong. 35.

Azeez M (2018) Bamboo, it's Chemical modification and products.

Cajee L (2018) Diversity of Bamboo species and it's utilization in the North-Eastern Region of India. International Journal for Research in Applied Science and Engineering Technology 6.

Dash S, Devi R and Kumar S (2020) Bamboos of Odisha: Socio-Medico-Economic Food wealth. Journal of Biodiversity and conservation. 4(1): 253-253.

Hammond KO (2006) Ghana: Cultivate bamboo for employment and income generation, Ghanaian Chronile(Accra)

Hossain MF, Islam MA and Numan SM (2015) Multiurpose Uses of Bamboo plants: A Review: International Research Journal of Biological Sciences. 4(12): 57-60.

Hunter IR (2003) Bamboo resources, uses and trade: the future?. Journal of Bamboo and Rattan. 2(4).319-326.

Patra R, Patnaik S and Mohapatra MD (2008) Issues and Challenges in management and trade of Bamboo in Orissa. RCDC Centre for Forestry and Governance, BBSR.

Janssen J (2000) INBAR Technical Report 20: Designing and building with Bamboo. Beijing: INBAR.

Kaminski S, Trujillo D and Lawrence A (2016) Structural Use of Bamboo: Part 1: Introduction to bamboo; The Structural Engineer.

Khan K and Hemalatha E (2015) A review on the Genus *Bambusa* and one particular species *Bambusa vulgaris* in Sabah(Malaysia); International Research Journal of Pharmacy 6(9): 580-584.

Liana A, Sumardi I and Daryono B (2017) Ethnobotany of bamboo in Sangirese, North Celebes. Bioscientifika: Journal of Biology anf Biology Education 9(1): 81-88.

Lobovikvo M and Paudel S (2007) World bamboo resources: A thematic study prepared in the frame work of the Global Forest Resources assessment. Non-wood forest products-18, Food and Agriculture Organization of the United Nations, Rome.

Nath AJ and Das AK (2008) Bamboo resources in the home gardens of Assam: A case study from Barak valley. Journal of Tropical Agriulture 46(1-2): 58-61.

Nirala D, Ambasta N and Kumari P (2017) A Review on Uses of Bamboo Including Ethno Botanical Importance. International Journal of Pure and Applied Bioscience 5(5): 515-523

Pradhan S (2015) Bamboo shoots: Beneficial effects on health 10: 78-81.

Rath B (2005) Bamboo in Orissa: Trade and Livelihood Perspective

Satya S, Singhal P, Bal LM and Sudhakar P (2012) Bamboo shoot: A potential source of food security. Mediterranean Journal of Nutrition and Metabolism 5(1): 1-10.

Sharma T and Borthakur S (2008) Ethnobotanical observations on Bamboos among Adi tribes in Arunachal Pradesh. Indian Journal of Traditional Knowledge 7(4): 594-597.

Shukla R, Sajal S, Diwedi P and Mishra A (2012) Medicinal Importance of Bamboo; International Journal of Bioharm and Phytochemical Research 1(1): 9-15.

Wong KM (2004) Bamboo: The Amazing Grass, A guide to the diversity and study of Bamboos in South-East Asia; IPGRI and University of Malaysia.

Index

D

E

F

G

H

I

K

L

M

N

O

P

R

S

T

U

V